ACROSS THE TABLE

ACROSS THE TABLE

AN **INTERNATIONAL OIL NEGOTIATOR** NAVIGATES
THE CHOPPY WATERS OF GLOBAL INTRIGUE

WILLIAM A. YOUNG

RIVER GROVE
BOOKS

This publication is designed to provide accurate and authoritative information in regard to the subject matter covered. It is sold with the understanding that the publisher and author are not engaged in rendering legal, accounting, or other professional services. If legal advice or other expert assistance is required, the services of a competent professional should be sought.

Published by River Grove Books
Austin, TX
www.rivergrovebooks.com

Copyright ©2015 William A. Young

All rights reserved.

No part of this book may be reproduced, stored in a retrieval system, or transmitted by any means, electronic, mechanical, photocopying, recording, or otherwise, without written permission from the copyright holder.

Distributed by River Grove Books

Design and composition by Greenleaf Book Group and Kim Lance
Cover design by Greenleaf Book Group and Kim Lance
Cover image: ©Thinkstock.com/sky_max; ©Thinkstock.com/Rawpixel Ltd

Cataloging-in-Publication data is available.

Print ISBN: 978-1-63299-041-9

eBook ISBN: 978-1-63299-042-6

First Edition

With loving gratitude to my family, friends, and business colleagues

Contents

Introduction / 1

ONE
Brinksmanship in Baku / 5

TWO
Surprise! "You Must Defend Your Proposal on
Live Nationwide Soviet TV" / 21

THREE
Summoned to the President's Office / 29

FOUR
The Not-So-Silky Road to Unitization / 43

FIVE
Blind Alleys and Threats / 51

SIX
Dangling the Lure of an Oil Export Pipeline through Georgia / 59

SEVEN
In Pursuit of Natural Gas off Russia's Sakhalin Island / 65

EIGHT
The "Wild West" Comes to Moscow / 79

NINE
Oil or Nothing: Opportunity Knocks East of the Urals / 97

TEN
Intersecting Destinies: A Near Miss in London / 113

ELEVEN
It Takes Two to Tangle: Vietnam during the US-Led Embargo / 117

TWELVE
Sea Slugs, Mao Tai, and Intense Negotiations / 135

THIRTEEN
Negotiating for a PTA Project in Guangdong / 145

FOURTEEN
Surrounded / 151

FIFTEEN
To the Ends of the Earth: Following in Rockefeller's Footsteps / 155

SIXTEEN
Concluding a Deal amid Philippine Chaos / 169

SEVENTEEN
An Imperfect Union in Buenos Aires / 177

EIGHTEEN
In the Land of the Pharaohs and Beyond / 185

NINETEEN
The Only Constant in the Negotiating Environment Is Change / 203

APPENDIX
Negotiations Principles: A Reference Guide / 207

Notes / 229

About the Author / 237

Introduction

INTERNATIONAL NEGOTIATIONS, ESPECIALLY THOSE IN THE oil and gas industry, can be exhilarating. Negotiators encounter all sorts of personalities across the table. They frequently travel to far-flung, unusual, and exotic locales. Some are very pleasant; some less so. The subject matter of the discussions (exploration for and development of petroleum, together with its transportation and marketing) is vital, both to the host governments and to the international energy companies. As a result, negotiators occasionally find themselves in pivotal conversations, as I did, with key international figures, including heads of state.

During a career that has spanned more than thirty years, I have held a wide range of executive, commercial, and financial positions with a handful of energy companies, including Amoco, BP, Burlington Resources, ConocoPhillips, and BG Group. Arguably, however, the most interesting and satisfying assignments have been those as an international negotiator. Those years also happened to coincide with some riveting moments in history in places like Russia, China, the Philippines, and Azerbaijan, to name a few. I consider myself very fortunate to have been entrusted with leading those negotiations and to have witnessed such fascinating world events swirling around me. Indeed, I have had the distinct privilege of playing a key role in negotiating and concluding oil and gas agreements around the world, including one that Azerbaijani President Heydar Aliyev called the "Deal of the Century."[1]

As you will see in the pages ahead, *Across the Table* brings international negotiations alive in a unique way. It is neither a pure negotiations text nor just a compilation of real-life anecdotes gleaned from my career. What I have done instead is to capture, in as lively a way as possible, what actually happened during my travels and negotiations, and then follow each episode with a short section that clarifies and elaborates upon the principles and learnings that can be derived from them. I believe one aspect informs the other, and it is my hope that the result is a richer and more fulfilling experience for you, the reader. In addition, for those of you who may find this book useful as a reference, I have created an appendix in which you will find the negotiations principles grouped together in outline form.

To make the book more engaging, I have opted to focus on the highlights of my business adventures by sometimes collapsing several trips into one or altering their sequence. In some cases, I have adopted fictitious names—where the individuals are not well known—to protect their privacy. Those caveats aside, I have sought to maintain the accuracy of the tales, subject only to the quality and clarity of my recollections and notes.

Clearly, international negotiations do not occur in a vacuum but rather are set in the context of the intrigue and complexities that geopolitical realities introduce. And "the other side" comes to the negotiating table with its own set of objectives, aspirations, and hidden agendas. Sometimes stark differences in culture, language, and values further complicate the picture. In addition, there typically is some degree of time pressure to add some zest to the mix. Both the host governments and corporate executives want the petroleum agreements signed yesterday!

I also have been fortunate to have had great colleagues along on these adventures. Negotiating, of course, is truly a "team sport," and I would be terribly remiss not to recognize the importance of the contributions of the experts on my team. Their insights have been instrumental in the joint successes we have achieved, and I will be forever grateful to them.

While the anecdotes in *Across the Table* are drawn from the international energy business, the lessons apply equally well to *all* business endeavors and indeed to our everyday lives. Whether we are debating with a friend or spouse the choice of a movie or restaurant, submitting an offer (and subsequent counteroffers) to purchase a house, or talking to neighbors about how to split equitably the costs of a fence on our shared property line, negotiations confront us at every turn. Of course, you can make a conscious choice not to negotiate by paying the asking price for a car or a house. To do so, however, undoubtedly means shelling out a significant premium over the true market value of such assets. It also means foregoing that sense of fulfillment that follows a successful negotiation. By that I do not mean vanquishing your opponent, of course, but rather the joy of identifying and agreeing upon an amicable and mutually acceptable path forward—one that benefits both parties.

It seems to me that there are two distinct groups of people in the world—those who relish the opportunity to haggle and those who find such encounters distasteful for various reasons. Personally, I view negotiations as both a challenge and an opportunity. However, I have noticed that even in business settings, managers are sometimes thrust into negotiations with little or no prior experience or training. That approach is akin to tossing a non-swimmer into the deep end of the pool and shouting from the water's edge, "Just swim!" Although some personalities are not naturally well suited to negotiations, most of us can reduce our stress levels by learning some basic principles, adhering to them, and preparing adequately in advance.

As with most endeavors in life, the skill of negotiations definitely benefits from practice. I certainly have made my share of mistakes over time but have learned from each. It is my hope that *Across the Table* will help you to avoid many common negotiations pitfalls. Enjoy!

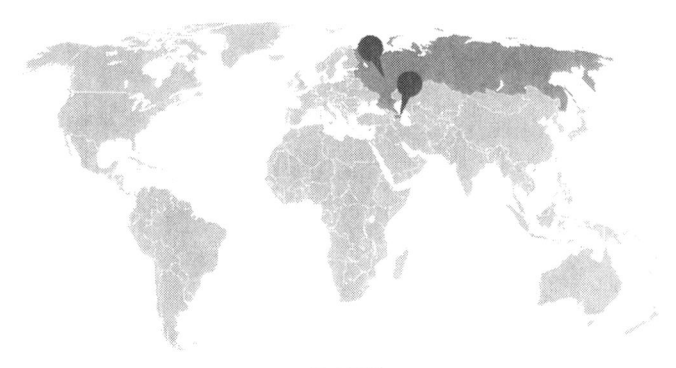

ONE

Brinksmanship in Baku

MOSCOW IN THE WINTER RESEMBLED A BLACK-AND-WHITE film, drab with little color. The snow along the city streets had long since been sullied by layers of charcoal gray road grit and by the belching exhaust of passing cars and trucks. I liked to think that there were only three possible scenarios on any given day in the Russian capital during the winter—it was either *snowing*, had *just* snowed, or was *about* to snow. A fresh luster of snow was a blessing since it would conceal, at least temporarily, some of the less aesthetically pleasing aspects of the city, and it also meant normally that the temperature was edging upward toward the freezing point. No matter how many layers of clothing one wore during the coldest days of the winter months, Moscow's unrelenting and cruel wind penetrated them with ease, like a knife slicing through butter.

I had always been fascinated by the Soviet Union, a society that represented in so many ways a mirror image of my own country, the United States. The USSR was ostensibly still a superpower in the late 1980s, though the cracks in its foundation were readily apparent, and it was soon to disintegrate into a number of autonomous nation-states, some of which remained quite dependent upon, and closely aligned with, Moscow. The people of the Soviet Union were generally very friendly. Muscovites, in particular, seemed so much like their Western counterparts. They were unfortunate victims of a system that neither encouraged nor rewarded risk taking and that failed to allocate resources—financial, human, and material—efficiently using the laws of supply and demand. The opportunity to learn more about the culture and history of the Soviet Union was a powerful lure, however, and I jumped at the chance to work on a project in Azerbaijan when the opportunity presented itself in late 1990.

At the time, I was vice president of negotiations for Amoco Caspian, the subsidiary of Chicago-based Amoco Corporation established specifically to work in Azerbaijan. I was to lead any negotiations which resulted from our business development efforts there. My predecessor had recently left Amoco, and I gladly accepted the chance to fill his shoes.

The USSR was a vast country with abundant natural resources. Azerbaijan, situated on the west coast of the landlocked Caspian Sea, had long been a regional hub for oil and gas production. Amoco was one of the more active companies in the Soviet Union at the time and had a handful of teams pursuing projects in widely dispersed sections of the country. Some were pure exploration plays, but most were less risky development opportunities with considerable exploration upside. Although a number of large oil and gas fields already had been discovered and delineated, the Soviets could only proceed with full-scale development if they could attract the requisite foreign investment and technical expertise. It was recognized that a great deal of front-end work would be required to launch any of these projects. Both petroleum laws and commercial understanding were embryonic at best and in many cases altogether lacking. On the other hand, the financial rewards could be correspondingly great. Of course, in the case of the Soviet Union, given the sizeable political and other risks involved, some of us asked ourselves now and again if being "fast followers" instead of leaders might be the wiser course of action. The decision, though, was "full speed ahead."

The USSR and Azerbaijan oil and gas ministries were jointly

sponsoring a competitive tender (auction) to award to a company—or consortium of companies—the rights to develop the Azeri oil field located offshore from the Apsheron Peninsula north of the Azerbaijan capital of Baku. The field had been discovered about a decade earlier but had not been developed, since the authorities lacked both the financial capital and technical capability to undertake this task on their own in these water depths.

Our mission in Baku was to purchase a data package that contained key information, both commercial and technical (e.g., seismic data, geologic maps, and well logs), upon which to base a competitive bid. Amoco was eager to get a toehold in the Azerbaijan sector of the Caspian, which was thought to contain substantial additional hydrocarbon potential. Modern-day drilling and production technology likely would result also in recovery of a larger percentage of the oil in place.

So, in January 1991, I made my first journey to the USSR since a holiday tour in the mid-'70s. The current trip also marked my inaugural visit to Baku, Azerbaijan, known as the "Paris of the Caucasus." I was accompanied by the head of Eurasian New Ventures (a geologist by training), a lawyer, and a geophysicist, who also doubled as our interpreter. We flew to Sheremetyevo II Airport in Moscow and then traveled the 52 kilometers (about 31 miles) by minivan to Vnukovo, an aviation facility which, at the time, primarily served destinations to the south. Vnukovo was located roughly due south of Sheremetyevo Airport and southwest of central Moscow.

Massive apartment blocks, many constructed during Stalin's reign and containing myriad virtually identical cramped flats, dotted our route. These structures were not pretty, but as if to compensate, the landscape was punctuated as well by the occasional Russian Orthodox church, with its bulbous steeple topped with a gold cross glistening in the sunlight. I reminded myself that many of the churches in Russia had been converted to museums, as the state did not recognize religion.

As we sped along, I reflected back on my 1975 trip to Russia when, as a US Naval officer, I had participated in an American Express tour of Moscow and Leningrad, as St. Petersburg was then known. Those were heady days when the period of détente had signaled a thaw in the icy cold relations between the US and the Soviet Union. The two countries entered into the Strategic Arms Limitation Treaty II (SALT II) talks and began a cooperative effort in space that culminated with the docking of two Apollo and Soyuz spacecraft in Earth orbit in July 1975.[1]

Still, there were the reminders of oppression back then—basic freedoms denied by a government preoccupied with maintaining iron-fisted control over its population. During my visit, I had observed an old lady, probably in her mid to late '70s, laying a small bunch of flowers at an exterior altar adjacent to a church. I started to photograph her as she knelt, but she spotted me, arose, and abruptly ran off down an alley, apparently fearing possible reprisal from the authorities should she later be identified as having prayed in public. I felt terrible for having prompted an elderly lady to flee from the church.

As it happened, my tour of Russia coincided with the celebration in Moscow of the thirtieth anniversary of VE Day (Victory in Europe). The Soviets played a vital role in World War II, forcing Nazi Germany to fight a two-front war. However, the Russians paid a heavy price, sustaining nearly twenty-four million deaths among its military and civilian ranks, over fifteen times as many as the three other Allies (US, UK, and France) combined.[2] As with most key anniversaries in Moscow, the obligatory parade of tanks and other weaponry in Red Square was intended to demonstrate the country's military might. Drilling nearby, in final preparation for their participation in the parade, were legions of children from the Soviet Young Pioneers organization (the equivalent of scouting), their necks adorned with bright red scarves.

The USSR felt like a parallel universe, yet one entirely separate from the rest of the world. The average Russian could not travel abroad and had only fleeting glimpses of what life must be like outside the Iron Curtain, such as when a recording of the Beatles would surface periodically. Though I suspected that they likely had no clue what the lyrics meant, young Russians would delight in singing along to "She Loves You" as they danced into the wee hours of the morning at nightclubs choked with thick smoke from their unfiltered cigarettes. Heavy drinking was rampant, especially among the male population. In a recent study published in the *British Journal of Ophthalmology*, of the 51,000 Russian males studied, 25 percent failed to live beyond age fifty-five.[3] I can only imagine that depression was widespread as well and largely left untreated. Nobody smiled much, perhaps a sign of how tough and unfulfilling their lives were.

English fluency was limited to those who had direct contact with Westerners, such as Intourist guides who led all tour groups once they set foot on Soviet soil. Foreign tour providers, such as American Express Travel, had no choice but to hand over the reins of control of their tours to Intourist upon arrival in Russia. The Intourist-operated hotels were almost certainly bugged, and the Soviets would have been well aware that I was a US Naval officer even though I had listed my profession generically as "office worker" on my visa application. I shared a room with another officer, and we were careful to avoid discussing anything classified or that might otherwise be of value to the Soviet regime.

The storefronts had glass windows, but that was where any similarity with Western retail outlets ended. Merchants would typically stack identical boxes in a pyramid in the window. Marketing was completely unnecessary, as consumers had few choices. Food was basic—meat and potatoes mainly, and of course borscht, the ubiquitous Russian soup containing beets as its primary ingredient. Soviet ice cream was delicious, but flavors were generally limited to vanilla and chocolate. Moscow restaurants, denoted by the Russian word *pectopah*, offered a less-than-inviting exterior. At night, most streets were dimly lit, surprising for a city of its size and influence. The inside of the restaurants was not much better. I recall dining in a dingy Georgian restaurant, ordering from a menu with no pictures and only the Cyrillic characters of the Russian language. No translation was available and nobody spoke any English. Accordingly, my selections were purely random, and when the lukewarm food arrived at my table, I still could not identify what I was about to consume.

A few everyday items were highly prized in Moscow. In the mid-1970s, one could exchange a mere stick of chewing gum for a handful of Soviet medals, and a pair of blue jeans was considered priceless. "Berioska shops" sold artwork and crafts to visiting tourists for hard currency.

Women did the same heavy work as men (for example, driving a bulldozer, which I witnessed firsthand), and the Russians proudly proclaimed that they were ahead of the West in "equal pay for equal work." They were probably right.

As I reflected on what I had observed on my 1975 trip, I was indeed curious how much the Soviet Union had changed in the roughly fifteen years since my previous visit. However, I did not have long to ponder this question. Jarring me abruptly out of my jet lag–induced daydream, a colleague in the front seat of the minivan broke the silence, pointing out a beautiful stand of white birch as we passed through a wooded stretch along Moscow's ring road.

"Yes, the birch trees are gorgeous," I freely admitted. As a teenager, I had enjoyed the stunning scenery in the movie *Dr. Zhivago*, based on Boris Pasternak's novel of the same name. Never mind that the movie was filmed in Canada rather than the Soviet Union. I tried to picture the birch with fresh green foliage in the springtime once all the dirty snowbanks had succumbed to the growing warmth of the sun and daylight had begun to stretch deep into the evening.

The ring road allowed us to make pretty good time, avoiding the traffic congestion of central Moscow. However, spring remained just a distant aspiration—a glimmer of hope—on this cold January afternoon. The sun hung low in the southern sky, and a gusty and bone-chilling northerly wind picked up litter and carried it across the highway. Men huddled along the side of the road, rubbing their hands together to keep warm, offering petrol for sale from large containers, a crude substitute (no pun intended) for a modern-day service station.

When we arrived at Vnukovo Airport, an antiquated and frankly rather dingy facility, we approached the Aeroflot counter where Svetlana, an attractive tall, blonde agent, informed us that the flight to Baku was already fully booked. We explained that we had important meetings in Baku the next morning and must travel there that evening in order not to miss them.

Understanding but apparently not appreciating our predicament, the stone-faced Svetlana inquired in perfect English but with a heavy Russian accent, "Will you be paying in hard currency?" She must have been saving her rare smiles for special occasions. Laboring behind the ticket counter at Vnukovo was clearly not one of those.

"*Da!*" we chimed in, practically in unison, as if we had rehearsed our response. When we assured the agent that we had come armed with US dollars, she indicated that she "would find us seats," to our collective relief and amazement. Airport security was almost nonexistent in those days, and we were able to drive our van right out onto the ice-covered tarmac where our bags were loaded onto the plane.

Later, after we had boarded the aircraft and wedged ourselves into the tiny rows of tattered and stained, cloth-covered seats, we stared out the window in disbelief as a uniformed airport official led a lady and two young children across the tarmac back to the terminal building. The realization hit us that our actions likely had precipitated the entirely unintended consequence of a Soviet woman and her family having to deplane and await a later flight. We were sickened by the thought, but it was too late realistically to do anything about it.

Our Tupolev-154 jet, stalwart of the Aeroflot fleet for decades, taxied and took off steeply into the twilight skies, banking sharply to the left. Once aloft, darkness quickly descended. The flight to Baku lacked any entertainment, unless one was to count people-watching, but at least it was also uneventful. About midway through the three-hour journey, the flight attendants circulated through the cabin to offer the passengers what appeared to be chicken broth. Since the crew had only enough plastic cups for about one-third of the travelers, they were forced to reuse the cups. I had previously observed Muscovites, while on my 1970s tour of the city, lining up during a heat wave to obtain water from a vending machine, using just one shared plastic cup. What an efficient way to spread illness! The prospect of sharing a plastic cup with the masses did not appeal then, and neither did it now. I looked up at the flight attendant and politely declined, "*Nyet, spasiba*."

A bit bored, I looked around to size up the passengers. Most of the men sported bushy black mustaches, had facial features bearing an uncanny resemblance to those of Iraqi President Saddam Hussein, and even wore the same beret-like caps. Iraq lay several hundred miles southwest of Baku where Operation Desert Storm, with the objective of liberating Kuwait from the grasp of the Iraqi dictator, had begun in earnest. Baku was also downwind of Iraq, should any chemical weapons be deployed by the Iraqi regime. I reminded myself that any such chemicals would have fully dispersed long before reaching Azerbaijan, but still the relative proximity of hostilities was slightly discomforting. As with the rest of the USSR, Azerbaijan did not officially recognize any religion, but the majority of its citizens were Muslim. I wondered where the Azeri sympathies lay in this conflict.

Baku was a fascinating city whose history was intertwined with the evolution and development of the international oil industry. Oil was recovered from surface diggings as early as the tenth century, and partly owing to deregulation of commercial extraction in 1872, production soared.[4] By the beginning of the twentieth century, roughly half the world's oil came from the Baku area.[5]

Upon our arrival at Baku's antiquated airport, we waited impatiently for about twenty minutes for our local agent. However, neither he nor his vehicle were anywhere in sight. It was now late evening, and we were exhausted. We had been traveling continuously since our departure from Houston the prior evening. It was clear that we needed to find an alternative form of transportation to reach our hotel.

Our interpreter, Peter, an American geophysicist from Serbia originally, convinced an ambulance driver to take us there. An emergency vehicle was certainly a creative, if somewhat inelegant, conveyance to our intended destination. I figured if we were involved in a crash, we were well-positioned for the trip to the hospital! I craned my neck to peer out the window to gain a first impression of my surroundings, but streetlights were few and a bit of mist had descended, reducing the visibility considerably.

As we approached the city limits, the ambulance slowed. Uniformed personnel from the Azeri army had established a roadblock. They motioned our vehicle to stop, ordered us out of the ambulance, and directed us to face the car and place our hands upon the roof. As they were all brandishing Soviet AK-47s, we readily complied. Baku was the site of sporadic violence, a spillover from the ongoing hostilities between Azeris and Armenians who were fighting for control of the disputed province of Nagorno-Karabakh. Military checkpoints had been established around the periphery of the capital city.

My understanding of the Russian language, though improving, was still rather rudimentary at the time. However, it was apparent that Peter was having real difficulty explaining why we were relying upon an ambulance for public transport. In fact, it was clear that the conversation was rapidly descending into a full-fledged altercation. The burly officers, grim faced, demanded to see our passports and visas. As I stood there with my back to the business end of their weapons, I pondered the wisdom of having volunteered to work in the Soviet Union when places like Western Europe, while perhaps less exciting, were considerably safer.

Then suddenly, just as quickly as the dispute arose, it was defused. The officers broke into broad smiles, returned our passports to us, and almost in a stage whisper muttered "*khorosho*" ("good"). *What could Peter have done or said?* I thought. We quickly piled back into the vehicle before they could change their minds. This incident served as a stark reminder that business trips to places like Baku carried with them certain attendant risks.

The capital at the time had little infrastructure and what did exist was in a state of general disrepair. It was clear that the Soviets had made a conscious decision not to reinvest in Azerbaijan, and trade was a one-way proposition (Baku to Moscow).

We based ourselves at the Old Intourist Hotel, which offered rooms with showers, at least in theory, though the spigots had an annoying habit of spewing out rather foul-smelling brown water. We were never quite sure if we were cleaner before or after bathing! Each room was identical, containing two single beds separated by a small night table. While not terribly glamorous, it did provide shelter and warmth. Moreover, the Old Intourist was considered one of the more secure and centrally situated hotels. It also represented a vast improvement over the Hotel Moskva, where I stayed on a subsequent trip. The rooms in that facility had a pervasive sewer odor, making it difficult to sleep, and harbored rather brazen and ubiquitous roaches. The Old Intourist was also located within an easy ten-minute stroll from where we would be meeting in the morning.

We traveled with a portable satellite system that encrypted our conversations so that we could communicate safely and securely with our home office. The technology was identical, I was told, to that used by television correspondents to file their reports during the Gulf War. It was surprisingly effective so long as one could lock on to the satellite, which hovered in geosynchronous orbit some 23 degrees above the southeastern horizon. Of course, one still had to take care in making calls since a naked speaking voice itself could easily be intercepted prior to encryption. The presumption was that the Old Intourist, while one of the few hotels with minimally acceptable facilities for Westerners (with emphasis on the word "minimally"!), was almost certainly bugged nonetheless.

And the Old Intourist was not without some security issues of its own. Later that night, I was awakened at about 2:30 a.m. by a lady's screaming in the car park directly below my second-story window. Wiping the sleep from my blurry eyes and looking somewhat disheveled, I am sure, I groped my way through the darkness to the lone window in my room and peered out. Illuminated by a street light about 100 yards away, two armed police or security guards—not sure which—were bludgeoning an elderly woman with their nightsticks. From my vantage point, she appeared to be at least sixty-five years old and certainly not at all menacing. She looked instead to be a poor peasant who was not putting up any resistance whatsoever. As there was clearly no option of dialing an operator to summon emergency assistance, I pulled on my trousers, ran down the corridor, and quickly descended the flight of stairs to the hotel lobby.

Given the hour, the lobby was entirely vacant and the desk unmanned. There was not even a night watchman. I strode over to the glass front door, all the while wondering what, if anything, I could do. As it turned out, no decision was required as both the woman and her tormentors had vanished into the night. I was left to surmise that the old lady had been dragged, kicking and screaming, into the back of an official vehicle and carried off to a location and fate unknown. I had neither any idea of what offense she stood accused nor why she was being beaten so savagely. I wondered what could have triggered the encounter, but to my mind there could be no reasonable explanation for the excessive force employed by the officers. The victim, in reality, did not seem to pose any threat, was apparently unarmed, and appeared completely defenseless. I was troubled that I had been unable to intervene in some way. For all the talk at the time of progress toward ending human rights abuses in the Soviet Union, it was clear to me that issues remained and that a segment of the population continued to suffer.

The next morning, our team had a brief meeting with Mr. Abassov, the president of Kaspmorneftegas (KMNG), the state-run production association that would be our partner in field development, should we be successful in the competitive tender process. In addition to a considerably larger office than his subordinates at KMNG, a glance at Abassov's desk made it clear that he was, in fact, the commander of the Baku operation. In front of him were no fewer than ten phones of various colors with a web of wires strung everywhere. It was rather humorous to watch Abassov leap into action whenever he received an incoming call. Knowing that he had limited time in which to connect to the caller, he would rather frantically pick up each phone in succession, yell "*da*," and then slam down each until ultimately he located the correct one. He did not seem the least bit embarrassed by what seemed to us a needlessly futile exercise. I struggled to contain my laughter as the situation reminded me of a late-night comedy sketch. Perhaps his assistant could recognize the phones by slightly different rings, but she was nowhere to be seen.

Abassov was an old, largely balding curmudgeon who, like many of his comrades, drank heavily. One of my colleagues remarked that the KMNG president had probably pickled his liver. Indeed, it appeared that the years had not been kind to him, and he looked considerably older than his actual chronological age. He died several years later. No doubt his excessive drinking was a major contributor and accelerator to his premature demise.

Upon conclusion of this introductory session, we moved down the hallway to a conference room where our conversations about the data package acquisition began in earnest. My mission, as the commercial negotiator, was to acquire the data parcel at a fair cost and for terms that would not expose our company to unreasonable risks or set adverse precedents for future negotiations. I had been given certain metes and bounds that limited my degrees of freedom in agreeing to the terms of the purchase.

I took a chair at the center of the long, rectangular wooden table, as is customary for the lead negotiator, flanked by my lawyer and the Eurasian New Ventures chief. Peter, our interpreter, positioned himself at the end of the table. KMNG's negotiator, Mr. Aliyev, seated across from me, dispensed with the customary exchange of pleasantries and immediately lodged a number of unreasonable demands. It was clear that the parties were nowhere close to agreeing upon the terms for Amoco to purchase the data set. We were, in fact, miles apart.

To make matters worse, I had the lawyer stage-whispering to me in one ear about how I could never accept the terms proposed by KMNG or "there would be hell to pay back in Houston," while somewhat distractingly I had the New Ventures chief admonishing me in my other ear that failure to return home with the data package would carry "very serious repercussions." I recall chuckling to myself about the negotiator being situated where he frequently is . . . at the center of controversy.

The hours dragged on as each side jockeyed for competitive advantage, but there was no discernible progress in the talks, no narrowing of the gap between our positions, and no helpful suggestions from either side as to how to get to a mutually satisfactory solution. We elected to call a brief recess to consider among ourselves the appropriate next move. After some lively internal debate, we returned to the table fifteen minutes later with a bottom-line take-it-or-leave-it offer. I was fully prepared to return to Houston empty-handed, if necessary, if the alternative was buyer's remorse. Accordingly, I realized that I would need to engage in the brinksmanship that such an ultimatum entails. I could feel my heart pounding, as I knew the stakes were high. Thankfully, this strategy paid off as my opposite number gave in on the key sticking points.

⊘ NEGOTIATIONS PRINCIPLES

One of the most important tools available to the negotiator is the option to walk away from a deal, or at the very least, to give a convincing performance of a willingness to do so. If the counterparty believes your instructions are to purchase the data package at "whatever cost," you can bet that you will pay far more than is actually necessary.

The willingness to walk away is at the very heart of any successful negotiation, but it is amazing to me how many businesspeople entrusted with leading a negotiation violate this fundamental principle. Some negotiators unwittingly send signals, either verbally or via body language, that they are determined to close the deal "come hell or high water." Perhaps it is because many negotiators are thrust into their roles with little or no training, or perhaps it is because their personalities in some cases are ill-suited to negotiations. Another contributing factor may be corporate bonus programs that incentivize negotiators to conclude specific transactions by a stipulated date (which sometimes is overly aggressive and unrealistic) with no mention of the quality of such deals. Whatever the reason, this sort of desperate behavior frequently results in a correspondingly desperate and materially adverse outcome. The other side in a negotiation can readily observe when a party is overly anxious to capture a deal, and this behavior inevitably translates into a poorer set of commercial terms.

Fortunately, in the instance of the Caspian data parcel, relatively few companies were interested in the offshore acreage, and the Azeris were eager to attract as many bidders to the competitive tender process as possible. I inferred that this was the nature of the competitive landscape since the Azeris seemed quite keen on luring Amoco to participate. If we did not purchase the data package, that would preclude submission of a subsequent bid in the tender process.

It is critical for the team to understand what the "breakpoints" or "show-stoppers" are for the negotiation. What are the "must have" provisions, and what is the maximum amount that should be paid? Again, if any member of the team displays body language that suggests that the bottom line is really not the final negotiating position, the success of the mission may be severely compromised.

In this case, I was also mindful that if we accepted terms for the data package that ran counter to our corporate requirements, we would be setting an adverse precedent for subsequent agreement negotiations from which it would be hard to escape. The other side would say, "Well, you willingly accepted Azeri courts for resolving any and all disputes arising from the data purchase agreement, so why are you now insisting upon international arbitration in London for the main exploration and production contract?"

It is also vital to see the discussion through the eyes of the counterparty. I suspected at the time that Aliyev would be in more hot water than I, had the discussion fallen through. (Actually, my instincts on this point might have been wrong. I discovered over time that in the Soviet Union there was generally far more risk for the counterparty in concluding a deal [which might later be criticized] than in either walking away or, more commonly, stalling.) I must say that I also harbored some fear that Aliyev did not appreciate how unreasonable his demands were, which could have created a real disconnect. Fortunately, everything came together.

I frequently reflect on this preliminary negotiation and ponder where Amoco (now assimilated into BP) would be today had it failed. My career as a commercial negotiator might also have been exceedingly short had I returned to Houston sans deal. The Caspian development project and the associated export pipeline would later become key assets in BP's international investment portfolio. At the time, I did not fully appreciate the history we were making. It was an early, very small, but very important step in what commentators would ultimately refer to as the "Deal of the Century."

We celebrated that night at Baku's "Chinese restaurant." I relished the thought of some Oriental food, as I was not particularly fond of Middle Eastern cuisine, which was all that was served at the Old Intourist and virtually every other eatery in town. Unfortunately, I was to be disappointed by this night out, as the term *Chinese restaurant* was clearly a misnomer. The establishment derived its name exclusively from the posters of the People's Republic of China that adorned its walls. Apparently, Oriental food was more an aspiration than a reality. In actuality, there was none to be had.

TWO

Surprise! "You Must Defend Your Proposal on Live Nationwide Soviet TV"

FOLLOWING THE NEGOTIATION OF THE DATA PARCEL PURchase, we returned to the US for several months of detailed analysis, development of a bidding strategy, and recommendations to management. Given the magnitude of the investments and the fact that a successful bid would entail new country entry, the proposal required the approval of Amoco Corporation's Board of Directors. The board gave its initial assent, subject to reviewing the final negotiated deal at a later date. I was responsible for preparing the management presentation and for contemporaneously producing a glossy booklet for the Soviet/Azeri bid committee, which contained the details of our proposal. The booklet also needed to portray convincingly our technical and financial qualifications as a world-class energy company that was more than capable of developing the Azeri field. The field was a worthy prize thought to contain billions of barrels of oil reserves. And all of this work was under significant time pressure as the deadline for bid submission loomed.

We gave some serious consideration to the factors that the bid committee might weigh most heavily in reaching its decision. A local consultant, whom the company had retained to provide actionable advice on how we might best present our bid, weighed in with his thoughts. He spoke of the need to develop strong interpersonal business relationships, especially important in this case since the Azeris had been "burned" so many times over their history and consequently tended not to trust outsiders. He further advised that we should show up with twenty-five sets of presentation handouts, which clearly spelled out the most compelling arguments in support of our bid. He foresaw a face-to-face meeting between the company's representatives and the bid committee, which comprised representatives of both the central USSR Oil & Gas Ministry and the Azeri government.

With our preparatory work completed, we boarded the corporate jet bound for Baku, making one overnight stop in Vienna for the pilots to rest. It was also a chance for the Amoco delegation to catch a delicious Viennese "last supper" before heading back into the thick of things in Baku. The "last supper" became a ritual of sorts for the Amoco team on each of our subsequent trips. The Gulfstream 3 was a great way to travel, cruising at an altitude above that employed by most commercial jets. It gave us the ability to touch down wherever we wanted, so long as we had first filed a flight plan. The downside on many trips, of course, was that senior management could, and inevitably did, drill us with questions throughout the flight, which made it less than a relaxing experience. If one chose to sleep, it was at his peril. Questions could be directed at groggy passengers at any point with the expectation that they would be fully prepared to respond. Still, in this case the corporate jet meant one fewer segment on Aeroflot, the Soviet air carrier. A flight on Aeroflot in those days always felt a bit like playing Russian roulette. Amoco was also becoming increasingly uneasy about its employees flying on Aeroflot. They questioned whether its maintenance procedures met minimal international standards. Rightly or wrongly, I pictured Aeroflot using duct tape to patch up its planes when the proper parts were unavailable!

The next day, as our plane descended to Baku, we gazed down at the landscape below. It was strewn with rusted oil derricks, and the sun shimmered off the accumulations of oil floating atop makeshift collection ponds. It was a very antiquated collection process. The crude would rise to the surface of the water where it could then be skimmed off and recovered. However, it was neither an aesthetically attractive nor an environmentally sensitive manner in which to produce oil. The old rigs were reminiscent of pictures I had seen of the earliest drilling in northwestern Pennsylvania in the mid-1800s. "Colonel" Edwin Drake drilled his famous exploration well to a depth of 69 ½ feet in Titusville, a quiet farming community, in 1859.[1] That well and those that followed triggered a worldwide oil boom and changed life for just about everyone on the planet.

Peering out the airplane window at the scene below, Mr. Blanton, president of Amoco Eurasia and head of our delegation, uttered aloud what I am sure others were quietly contemplating, "My God, what have you gotten us into?" Oil sheens were evident on just about every inland pond. Elsewhere were large stretches of exposed soil with little vegetation breaking up the otherwise drab, brown landscape.

From this altitude, the Oil Rocks field could be seen in the distance, extending out from the shoreline. The Azeris relied upon a series of causeways as an ingenious way to position oil wells throughout the shallow-water field. The Oil Rocks field had been producing for many decades. In fact, the first offshore well in the world was drilled not far from here off the Apsheron Peninsula in the late nineteenth and early twentieth centuries.[2] I was reminded of Azerbaijan's proud heritage as one of the world's most famous oil provinces. Baku, after all, was where the Nobel brothers had made their fortune, constructing a pipeline in the 1870s across Georgia to the port city of Batumi on the Black Sea. The script of the 1999 James Bond movie, *The World Is Not Enough*, appears to have been loosely based upon the export pipeline that runs from the Sangachal facility just south of Baku via the Georgian capital of Tbilisi to Ceyhan, Turkey, on the Mediterranean coast.[3]

We were met at the airport by our local representative, a couple of Azeri dignitaries, and several members of KMNG, one of whom proudly announced that Mr. Blanton would accompany its senior management on a hunting trip to a game preserve near the border with Iran. We had planned to stay together as a unit in order to prepare for the presentation that was to occur two days hence. However, there was simply no talking the Azeris out of this invitation without offending or insulting them. Since one of our key objectives was to establish a foundation of trust and to solidify the working relationship with the Azeris, we accepted their plans without reservation or hesitation.

This unexpected development was rather ironic since Mr. Blanton had absolutely no interest whatsoever in hunting. A dedicated and capable executive, "his work was his life and his life was his work," as the saying goes. He enjoyed immersing himself thoroughly in it. We watched wistfully as his jet-black Chaika, sandwiched between two security vehicles, disappeared into the distance. A Chaika, the Russian equivalent of a Cadillac, was normally reserved for transporting important state officials or other VIPs.

The real shock, however, came when one of the KMNG officials then turned to those of us who were left behind and mentioned almost offhandedly, "I guess you know that you will be defending your bid 'in the Soviet manner' before the bid committee on national television." "In the Soviet manner" was akin to defending one's PhD dissertation. Having come prepared for a meeting of perhaps as many as thirty-five people, we were dumfounded to learn that we were to appear live on Soviet television, which reached a huge audience across the nation's eleven time zones. Meanwhile, our leader, completely oblivious to this change of plans, had motored off to shoot whatever game still roamed the hunting preserve near the Iranian frontier!

Although I was successful in masking my emotions, I was inwardly seething with anger. I wondered why we had paid a local representative handsomely for advice and intelligence that had proven to be so totally erroneous. However, there was little time now for regrets or finger-pointing. We jumped into action, scuttling our original plans and writing what amounted to political speeches about why the Amoco bid would be in the best interest of the Azeri and Soviet people. This was no easy task as the Azeris, as noted earlier, were already highly suspicious of the motives of Western energy companies, having been taken advantage of on numerous occasions throughout their history, most recently by the USSR's production associations and ministries. Caspian oil had been exported to Russian refineries for years, but somehow the resulting revenues never found their way back to Baku to improve the lives of ordinary Azeris. The Azerbaijan capital was the oil hub of the Soviet Union, where most of the oil field expertise resided—the equivalent of Aberdeen in the UK or Houston in the US. As recently as 1941, Azerbaijan produced 23.5 million tons of oil, nearly three-quarters of the Soviet Union's total production.[4] Sadly, as a result of the aforementioned imbalanced trade flow, the Azeris had precious little to show for their efforts.

As promised, on the day of the presentation, we were greeted by the bright glare of television lights and cameras at the venue in a large municipal auditorium filled to capacity. Thankfully, in the interim, we had reunited with Mr. Blanton upon his return from the Iranian frontier and agreed upon a plan for the presentation and subsequent question and answer session. After Mr. Blanton made some initial remarks, the task would fall to four of us to present the main body of the presentation—exploration, development/production, commercial/legal, and economic. I spoke of a "chicken in every oven" and the multiplier effect (an economic concept that contends that for every dollar spent in direct investment, a multiple of three to four dollars would be generated in spin-off enterprises or related businesses, ranging from industry subcontractors to the local barber and butcher). The presentation itself proceeded flawlessly, though it was tricky to judge an audience, especially one hailing from a vastly different culture, until questions were put forward at the conclusion. We hoped we would then be in a better position to read whether or not they understood the presentation and were favorably disposed to our proposal.

Indeed, the most difficult aspect lay ahead—the grilling by the Bid Committee. Its largely gray-haired members, clad in business suits, were positioned at a long table on the floor between the stage and the public seating. Would these venerable experts welcome Amoco's involvement in their industry, or would they see us as superfluous, meddlesome, or intrusive? Would we be viewed as Western interlopers with self-serving objectives? It proved difficult to get a fix on the panel, as most of the representatives sat there patiently but entirely expressionless. One, I noticed, had even drifted off to sleep. I hoped that it was not my scintillating presentation that had sent him over the edge into never-never land!

Mr. Blanton had requested that all the questions be held for the Q&A session following the formal presentation. We then would answer them one by one. As each query came in, Mr. Blanton would scrawl it on a small slip of paper and pass it along to the team member he expected to field that response once all the questions had been submitted. This technique proved highly effective since it allowed sufficient time for each of us to gather our thoughts and to develop the best arguments to put forward. Russian/English interpretation was available throughout the meeting, which also provided us additional time to think. Moreover, Amoco had the good fortune of being the third and final company to present its bid. Since each of the presentations was open to the public, we were able to observe and improve upon the techniques used by our two competitors, BP and Unocal, when they made their pitches on the previous day. We felt that we had handled the questions rather well. We eyed the bid committee cautiously as we filed out of the auditorium, but they provided no hint—verbal or otherwise—as to which of the three companies had made the best impression.

We celebrated what we hoped was a successful bid presentation that evening by dining at the famed Caravan Serai Restaurant in Baku. Baku was one of the stops on the Silk Road. For centuries, caravans would stop here, their camels would be watered, and the owners would lodge for the night.[5] They would feast on lamb and other delicacies before continuing their long and arduous overland journey to China. The Caravan Serai offered a number of grilled lamb dishes, stews, and Middle Eastern specialties.

I recall that on a subsequent occasion, when we were joined by our KMNG hosts, I unwittingly ordered what one of my colleagues fittingly called "aorta stew." This lamb concoction, immersed in thick brown gravy, was tasty enough but seemed to use those parts of the animal that we in the West would customarily discard, the aorta being prime among them. I played with the lamb for a bit with my fork but then decided to let it sit on my plate until the waiter could come by to collect it. The highlight, or perhaps the lowlight, occurred after the stew had cooled. When the waiter arrived at my place, he proceeded first to pick up my fork. Mired in congealed fat, the fork momentarily lifted the entire plate with it before the platter came crashing back down to the table. Yuck! And I attracted the very attention I was hoping to avoid.

Of course, all of our meals in Baku at which our KMNG hosts were present were drowned in vodka toasts. They drank their alcohol swiftly and with no dilutants. The most "deadly" toast was the dreaded third one, which was dedicated to "the women—wives, girlfriends, mothers, and daughters—that we had left behind." It entailed downing the entire glass of vodka in one long continuous burning swallow. In the Soviet Union, I learned with time that those who would not join in the toasting process were regarded with suspicion and distrust. Of necessity, I gradually developed surreptitious methods to water down the vodka that, so far as I know, escaped detection. Mercifully, I survived to negotiate another day, though perhaps with a somewhat curtailed lifespan!

⊘ NEGOTIATIONS PRINCIPLES

An important lesson from this episode is that, regardless of the cultural backdrop, business is always transacted on a foundation of trust. Trust was understandably in short supply in Baku in the early days, when only our word and our company's proven public track record stood as evidence of why we should be counted upon by the Azeri leaders to build a brighter future for their people. It would take years before that trust was fully established, which may explain why, in part, what became known as the "Deal of the Century" took years to complete. Trust gradually accumulated on the back of a series of one-on-one professional business relationships. The difficulty (and importance) of developing this working rapport cannot be overexaggerated, as in many cases the two sides culturally had so little in common.

A second principle which underpinned our success in Baku at the bid presentation was the importance of doing our homework/preparation in advance of the trip and remaining unflustered in the face of whatever questions or comments came our way. Indeed, we were prepared for just about any eventuality, which was good since we certainly were not expecting the huge TV audience we reached. I am convinced that reworking our speeches to focus on the value of field development to the ordinary man/woman was a vast improvement over the alternative of sticking stubbornly to a more conventional business presentation and script, which might have appealed to some on the bid committee but had little impact on others and the general citizenry. Of course, we needed to and did target both audiences. The bid committee itself was comprised of equal numbers of representatives from Moscow and Baku, meaning that our messages also needed to be carefully crafted to resonate both with local officials and those from the central government.

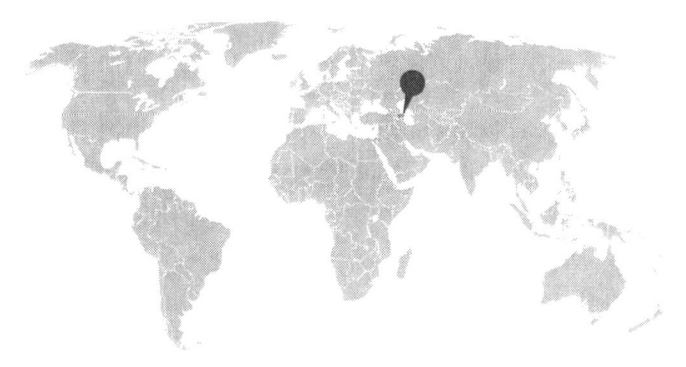

THREE

Summoned to the President's Office

FOR THE TENDER COMPETITION, AMOCO TEAMED UP WITH McDermott, a well-known engineering firm that had been active for several years in the Caspian. Its knowledge of the Azerbaijan market, including what oil field equipment and services were available locally and what needed to be imported, proved quite helpful.

About a month later we learned that the Amoco/McDermott alliance had "won the competition." Of course, our victory only really entitled us to the exclusive right to negotiate with KMNG for some finite but unspecified period. We really had no idea how much time we would be allowed, or for that matter, how much time it would take to negotiate a deal. However, we suspected that the Azeris would be impatient, wishing to monetize their oil reserves as soon as possible. In reality, a number of Azeris expressed a preference to keep the oil in the ground "for future generations." I never really understood that thinking, but I suppose it was a byproduct of the Russians "stealing their oil wealth" in the past.

The discussions with KMNG proved difficult and protracted. Amoco wound up inviting both BP and Los Angeles–based Unocal (subsequently acquired by Chevron) to participate in its consortium, subject to government approval, which brings to mind another anecdote.

On one of our first negotiating trips to Baku, in which representatives from all four primary companies (Amoco, BP, Unocal, and McDermott) participated, I was midway through leading the morning discussions at the Old Intourist Hotel, which also doubled as Amoco's local office, when there was a knock at the door. It was Natig Aliyev, personal aide to Azerbaijan President Mutalibov, who came up to me, leaned over, and whispered, "The president would like to see you right now. Please come alone. Do not bring an interpreter." I adjourned the meeting, scurried down the stairs with Natig, out the front door, and into a jet-black Chaika parked outside. We sped off to the presidential palace.

Azerbaijan President Mutalibov at the palace

Upon our arrival, President Mutalibov greeted me, his comments being interpreted by his minister of foreign relations, who spoke English flawlessly. The president's gaze fixated on me as if to size me up. As he beckoned me into the inner sanctum of his office (Natig waited outside), the president again glanced at me and continued, "You look tired. Would you like to visit one of our sanatoriums?"

My discomfort at the president's suggestion was no doubt plainly evident on my face. I had a wife and two children at home and wanted to see them again! (In the former Soviet Union, sanatoriums were more than just places to rest and retool; they could also be full-fledged medical facilities for those needing to convalesce. I did not want the state determining what treatment I might need!) It was unclear whether the offer had been advanced as a hospitable gesture of goodwill or as a thinly veiled threat. Conservatively choosing to believe it was the latter, I quickly, forcefully, but politely declined. *Did I really look that tired?* I wondered.

Mutalibov then shifted seamlessly to the primary reason why he had sought the meeting. "I see you are planning to share some of your interest in the Azeri field development project with British Petroleum." I nodded affirmatively. He continued, "We really do not want the British involved in this project. I assume you know why?"

"No."

"Many years ago, the British executed twenty-six of our commissars, and we have never forgotten or forgiven. That is why the old name of this field was '26 Commissars Field.' So, I am sure you can appreciate why we don't want BP in this project." Shortly after assuming power in 1917, Lenin was engaged in an epic struggle with British and Turkish forces who were trying to prevent Communism being imposed upon Azerbaijan. When a pro-independence government briefly seized control in Baku, the twenty-six Bolshevik commissars from the prior regime, who were aligned with Lenin, were rounded up, arrested, and summarily executed. While it was unclear that the British were directly involved, the blame was laid at their doorstep nevertheless. Shortly thereafter, the Communists regained control, and the twenty-six commissars were hailed as martyrs for the Bolshevik cause.[1]

I found myself in the rather odd position of arguing the merits of BP, whom we had just defeated in the tender competition, and explaining what benefits and strengths the British company would bring to the consortium. After considerable back and forth, the president agreed to let BP have an interest but at a diminished participation level compared to the one the two companies had previously discussed and conditionally agreed.

The president then explained, at least in part, why the Azerbaijan representatives on the bid committee had voted for Amoco and McDermott in the first place. "You are an American company, as are Unocal and McDermott. We feel you can help us convince the United States government that our claims to Nagorno-Karabakh are legitimate and right. The Armenian lobby is very strong in the US. Perhaps that is because in World War II an Armenian doctor helped [former senator Bob] Dole recover from his war injuries."

President Mutalibov continued, "In addition, America favors Armenia because they are *Christian* and we are *Muslim*. We believe your company's influence with the US government could be very valuable to us."

I then launched into an explanation of the reasons why Amoco, as an international energy company, did not participate in political discourse, much less take sides in a border conflict. In fact, the company could not represent a foreign government's interests without registering itself as a foreign agent, a step that we were not prepared to take. The president reluctantly accepted the point, though reading his expression I fully expected him to raise the topic again at a later date and in a different venue. Equally, he could see that I was not going to budge on the issue.

The meeting ended as abruptly as it began. I thanked the president for his flexibility with regard to BP and grabbed my coat. Natig, who had been waiting in the outer chamber, chauffeured me back to the Old Intourist Hotel. As we descended the steep hill from the palace, he was anxious to hear how the meeting went. I replied simply that all was fine without sharing that we had just agreed the interests that each company would have in the Azeri field development. I had no doubt that this topic would be a major bone of contention with BP, and I did not want the information leaked prematurely. Amoco would now have the extraordinarily difficult task of explaining to the London-based company why its interest in the project, despite my best efforts, had been materially diminished. They would not immediately appreciate that the government had wanted them excluded altogether. Queen Elizabeth II would not be happy!

> **PRESIDENT MUTALIBOV WAS** correct when he said that an Armenian doctor helped Dole recover from his war injuries. Dole was serving in the US military forces in Castel d'Aiano in the Italian Apennine Mountains at the time. As Dole recounted in his book, *One Soldier's Story: A Memoir*, "I didn't know it at the time, but whatever it was that hit me had ripped apart my shoulder, breaking my collarbone and my right arm, smashing down into my vertebrae, and damaging my spinal cord."[2] *Washington Post* reporter Jonathan Yardley, who reviewed Dole's book, said it was "a miracle that Dole survived the several hours before medics reached him, then the excruciating trip to the hospital, then the transatlantic flight, then the years of treatment, surgery and therapy."[3]
>
> A number of doctors operated on Dole, but it was Dr. Hampar Kelikian, an Armenian immigrant, who he indicated helped him most with his physical and psychological afflictions. Dole credits Dr. Kelikian with encouraging him to adopt a new, more positive attitude and to focus on what he had left and what he could do with it. Dr. Kelikian refused to accept payment for injured American GIs he helped.[4]

> ## ⊙ NEGOTIATIONS PRINCIPLES
>
> Negotiations do not play out in a vacuum, as was readily apparent in this episode. Geopolitical considerations and history were major factors both in the award of the block to Amoco/McDermott and in the stakes that the government was prepared to grant to companies within Amoco's consortium. The authorities clearly wanted an American company to lead the effort, and they were loath to allow a British firm to be involved. We knew the previous name of the field ("26 Commissars") but had not made the connection that historically the Soviets had blamed the British for the executions.
>
> Fortunately, prior to journeying to Baku, my colleagues and I had negotiated a Memorandum of Understanding (MOU) with BP and the other companies. Although the MOU spelled out what shareholding each party would have in the consortium, it also contained a provision that, should the Azerbaijan government reject the arrangement struck among the companies, the MOU would terminate.

When I returned to Houston, I learned that I had been appointed to deliver the distasteful message to BP's lead negotiator that the MOU had been terminated and that, despite my best efforts, the Azerbaijan president had reduced BP's stake in the consortium. No surprise—the British company's lead negotiator was outraged, believing Amoco had worked behind the scenes to undermine BP's position. In reality, nothing could be further from the truth.

That night, at a prearranged cocktail party for each of the companies' commercial and technical teams who were present in Houston at the time, BP's chief negotiator tracked down Mr. Blanton. He then proceeded to go "eyeball-to-eyeball" with Mr. Blanton, a Southern gentleman who was unaccustomed to being treated so disrespectfully, especially at such a public gathering and in front of his subordinates. Not a wise move! The BP negotiator's actions nearly caused Amoco to drop the British firm entirely. However, Amoco was reluctant to retain too much of the megaproject with all the risks that it entailed. As one of our senior executives later pointed out insightfully, "The Azeri project likely will be either a company-maker or a company-breaker."

This incident served to erode what little trust existed between Amoco and BP, and BP launched a stealth effort to secure for itself another field development project—Chirag. It would then be in a stronger position to compete directly with Amoco for the scarce oil field equipment and services available in the Caspian.

Mr. Blanton addressing an Azeri delegation through an interpreter

On subsequent trips to Baku, we continued to lodge at the Old Intourist Hotel. The food there and elsewhere in town was not only of poor quality but the selection limited and boring. Furthermore, both food handling and sanitation were concerns. On one occasion, our attorney Michael contracted food poisoning from consuming fresh cherries, either at the hotel or at the Caravanserai, as he later reconstructed events. He was so dehydrated and weak that we were forced to leave him behind in Baku with a British medic to nurse him back to health. Amoco had retained the medic for precisely this sort of eventuality. The attorney returned to the US a couple weeks later, looking drawn and a bit frail. I assumed that he was also less than enthusiastic about the prospect of returning to Azerbaijan anytime soon.

There would be periodic entertainment at the Old Intourist restaurant, a mixed blessing really. The music seemed rather discordant to our Western ears. One of my team would joke that funeral music had been popularized in Azerbaijan. I recall a wedding reception one evening at the hotel where the live music sounded particularly off-key and mournful. Our tax attorney, Robert, rather brazenly approached the bandleader and asked, "How much are you being paid to perform, comrade?" Upon hearing the response, he offered to double the amount for the group to stop playing. While the incident was amusing and was clearly intended only in jest, humor is not well understood and appreciated uniformly across differing cultures. I feared he might well have offended them. Fortunately, no riot ensued and we retreated to our hotel rooms for the night. As if to torment us, the music followed us around the hotel, as the walls were paper thin and easily penetrated. It seemed as if the band had elected to play even louder in protest.

There was not much to do with what little free time we had, but my boss chose on one visit to Baku to bathe in the Caspian. I had read far too much about pollution in the landlocked sea to join him. Pesticides, heavy metals, and oil residue were reportedly among the pollutants. While most of the highly valued caviar was obtained from sturgeon further north in the Caspian, it made me wonder how healthy the delicacy was to ingest on a regular basis. That said, I did consume my share, both in Baku and abroad.

We were still flying Aeroflot to and from Moscow as the most direct and reliable route to the Azerbaijan capital. One morning, as we boarded the plane back to Russia, the Aeroflot female flight attendant glanced at our attorney Michael, smiled slightly (a true rarity for Aeroflot staff in those days), and gushed, "Beautiful man!" Blushing, our attorney, rarely at a loss for words, was rendered speechless. Our tax attorney, Robert, straightening his tie, jumped into the conversation with his best Rodney Dangerfield impression, "What about me? I never get any respect!"

From my seat I was able to see into the cockpit since the door had swung open. The copilot was holding in his hand what looked like a partially consumed vodka bottle. "Surely that must be water!" I muttered under my breath. We had already begun to taxi, so I knew my fate rested squarely in the hands of the crew regardless of what beverage the copilot had decided to imbibe before takeoff and during the journey.

Like the Aeroflot flights between Baku and Moscow, the negotiations in ensuing months proved to be a bumpy ride, and the prize of a successfully negotiated agreement remained elusive. It soon became apparent that KMNG's lead negotiator, Mr. Aliyev, was unfamiliar with even the most basic aspects of oil agreements and economics. What was particularly surprising was that as we discussed the variables in an economic analysis of the project, such as price, production rate, capital expenditures, operating expense, and inflation, Aliyev stated that he was well aware that world crude prices were determined by a formula based upon depreciation. Repeated efforts to explain that crude oil price levels were actually a function of supply and demand in a global market made no headway, and this was just one example of the cultural and experiential chasm between the parties.

The Amoco group was proposing a production-sharing contract (PSC). In this arrangement, the Amoco consortium and KMNG would each be "Contractors" and would share equally the project expenditures and take the risks associated with the appraisal and development program. In return, the Contractors would be allowed to recover their costs from a portion of the available petroleum ("cost oil"). The remaining production would constitute "profit oil" to be shared by the Amoco group, KMNG, and the government of Azerbaijan as petroleum owner. The government would retain the vast majority of the profit oil, particularly later in the life of the project once the Contractors' investment had been recouped. The government would also pay the Contractors' taxes on their behalf. The proposal was designed to allow the Contractors to earn a reasonable, but not excessive, rate of return. Production-sharing contracts can be rather complex, even for those who are reasonably familiar with them, so it was no surprise that the concepts were alien to the KMNG team.

How were we going to find common ground to reach the goal of signing a PSC, especially when the level of mutual trust was still exceedingly low? If they did not grasp the concepts we were discussing, they understandably would be hesitant to sign a contract, and if they did, they might fear that they were being taken advantage of by the foreign oil companies (or FOCs, as they came to be known).

All the issues were not on the other side of the table, of course. The FOCs had their own difficulties in agreeing which topics to address first, which issues represented negotiation "breakpoints," and even the notion of one person acting as spokesperson at the table for a common position previously arrived at during internal consortium discussions. One of the geoscientists, who seemingly felt the best way forward was to share freely everything he knew with the other side, would also show up from time to time, complicating the discussions. While this open exchange of information might work well in technical circles where shared learning was the objective, it was usually less well suited to the conduct of effective commercial negotiations.

⊘ NEGOTIATIONS PRINCIPLES

When two parties negotiate, they need to have similar levels of understanding and "speak the same language." The Azeri team was comprised of bright people who had not been exposed to concepts and understandings that were widely shared and accepted throughout much of the world. We quickly and correctly concluded that it was not only in their interest, but also in ours, to have KMNG retain their own advisors who could represent them competently in discussions. This approach, we reasoned, would accelerate the negotiations process considerably, but the trick would be, "Who would pay for these advisors?" If the FOCs paid, the Azeris would rightly question the independence of the advisors and therefore the reliability of the recommendations they made, yet KMNG had no hard currency to compensate Western legal/commercial counsel.

As we researched this issue, we discovered that the World Bank had technical assistance funds that could be used for this purpose. A source of funding was accordingly identified, and KMNG's successor organization, the State Oil Company of Azerbaijan, retained the American law firm of Akin Gump Strauss Hauer & Feld LLP to represent them and provide advice and counsel during the negotiations. This step was a major breakthrough, allowing a common language of business to be spoken and understood on both sides of the table.

Another key negotiations principle is that agreements that reflect an underlying disproportionate and inequitable deal rarely last, nor should they. The Contractors were seeking a "fair return" for the risk they were taking and the sizeable investments they were making. At the same time, they knew that the Azerbaijan government, which would need to approve the PSC, and the Parliament, which would enact it into law, would require a clear economic benefit for Azerbaijan. The FOCs wanted the PSC to be passed by the Parliament so that subsequent administrations could not claim it was not adequately considered, debated, negotiated, and approved. If the current executive branch office holders were later discredited, this approach made subsequent disavowal of the PSC less likely and added a measure of stability.

Of course, not everyone's personality is well suited to negotiations. While "thinking aloud" and sharing all one knows may be beneficial in a scientific setting, it is probably not a wise approach at the negotiating table where the consortium is trying to adhere to a carefully crafted message. Honesty and transparency, of course, are critical pillars in any negotiation, but passing along more information than is required can compromise the interests of a party in a negotiation or confuse the main messages. For example, the fact that a party, if necessary, is prepared to spend twice as much to purchase an asset as he is proposing is clearly information that party needs to keep confidential.

It is also important that a negotiator have a certain degree of separation from the project that is the subject of the discussions. A person linked too closely to a venture may be emotionally invested in its success, making the willingness to walk away less likely and problematic. A project leader, for instance, who may have been nurturing the proposed venture from inception, would typically be a strong advocate whose bonus may be tied to its capture by a certain time deadline. Accordingly, such a person might be less inclined to fight tenaciously for optimal terms and therefore would not be an ideal candidate for the role of negotiator. In the same vein, it is generally not a good idea to have a negotiator reporting to the project leader. If a negotiator's performance appraisal is in the hands of that person, the negotiator's independence may be compromised. A far better structure is to maintain a pool of international negotiators that provide a service to a new ventures group but do not report directly to them. The quality of deals that are struck should be measured against well understood and accepted metrics.

Listening and observing are two underappreciated behaviors at the negotiating table. It is sometimes possible to pick up clues as to the other side's true underlying positions by "active listening" and observation of their body language. Equally, there is sometimes an advantage to be had in letting the other side make their case first and disclose their views in the process. For example, maybe their asking price for an asset is 75 percent of what you were going to propose, allowing you to pocket the savings.

Frequently, a negotiation is a "zero sum game," meaning that there is a fixed level of "economic rent" to be allocated between the parties. If I were to agree to provide more of the net cash flow to the government, then there would be less available for my company. That said, there are sometimes opportunities to expand the pool of economic rent to be shared so that both parties can incrementally benefit. We will talk later about disproportionate valuations in a negotiation, a situation where one party values an element more than the other party does. Such elements can sometimes be used to craft a compromise solution to narrow or close the gap between parties in a difficult negotiation.

FOUR

The Not-So-Silky Road to Unitization

WITH THE BREAKUP OF THE SOVIET UNION IN DECEMBER 1991, some of the officials with whom we were dealing disappeared virtually overnight. One of the persons who vanished was Azerbaijan's first secretary of the Communist party. The story circulating in Baku was that he had fled the country, reportedly taking refuge in Switzerland where he had allegedly squirreled away a significant sum of money in Swiss bank accounts. I never saw a confirmation of this report, so I must regard it as one of the many rumors circulating in the Azeri capital at the time, but the story certainly seemed plausible. Since I never laid eyes on him again, I have no idea what ultimately became of him.

A distinctly more optimistic feeling permeated the environment in Baku now that the yoke of the USSR had been thrown off. The KMNG employees talked exuberantly about the dawning of a new day. I overheard their chief geologist and a colleague rejoicing over the abandonment of a program that rewarded one person for turning in another to the authorities for any suspected violation of the law or threat against the state. In the old days, a person simply acting suspiciously could prompt a neighbor to pass his name along to the police. This program had clearly accomplished its intended goal of spreading mistrust among the population and ensuring that no grassroots movements could threaten the authority of the state.

Azerbaijan became an independent country on August 30, 1991, and decided to join the Commonwealth of Independent States (CIS), which was formed from many of the former Soviet republics.[1] In September 1992, with the issuance of Decree 200, the Azeri government established the State Oil Company of the Azerbaijan Republic (SOCAR). The new management structure was delineated in Decree 328, which was issued the following February.[2] KMNG was no more. Fortunately, however, many of the personnel with whom we had been dealing at KMNG reappeared in similar roles at SOCAR, providing for good continuity.

On the surface at least, relations between Amoco and BP seemed to have healed. The British company hosted a reception for the Azeri field consortium at their Hill Street office in London, followed by tea at the Ritz. It was at this reception that I met Lord John Browne for the first time. He was then head of BP's upstream businesses but would go on to become BP's group chief executive. Lord Browne had many interesting stories to recount, but it was difficult for me to hear him sometimes, given the background noise at the reception and the height difference between the rather diminutive Lord Browne and me (six foot, five inches, or 195 centimeters).

Our height differential reminds me of a humorous incident that occurred years later when I was working for BP in their Britannic House offices in the city of London. Definitely not a morning person by nature, I would reward myself each day as I arrived at work by taking the lift up one level from the ground floor to the first floor (equivalent to the second story in the US) where my desk was situated. BP utilized the "open plan" layout, which meant no office walls as such, and inevitably one came to know everything about everyone within a radius of 15 feet. At any other time of the day, I would willingly trudge up the dimly lit marble stairs of Britannic House, but when I was still groggy in the morning, the elevator ride was a luxury in which I happily indulged myself.

One misty gray morning I was running a bit late and scurried into the lift behind a short little man whose identity was not readily discernible from the rear, and pressed "1" on the control panel. As the gentleman turned around, I realized to my horror that it was Lord Browne. I am not sure if he recognized me or not, but his glare clearly suggested what he was thinking: "I am far more important than you, whoever you are. What are you doing riding the lift up just one floor when I have pressing matters to which I need to turn my attention?"

Thinking quickly on my feet, which rarely happens even at a more civilized hour of the day, let alone in the early morning before I have had my requisite dose of caffeine, I limped out of the elevator when the door slid open. I figured Lord Browne would be upset at himself for harboring negative feelings about someone who clearly was physically handicapped.

When I arrived at my work area, I proudly recounted the story to colleagues. One of them correctly pointed out, "You know that every time you see Browne from now on, you will be forced to limp."

"Damn! You're right!"

Suddenly, my spur-of-the-moment masterful plan was, well, not quite so masterful. *Oh well*, I thought, *it's only a career!*

Returning to the subject at hand, as negotiations on the Azeri field development ground on, sometimes circling with little discernible forward progress, BP was busily courting the Azeri government to award it the Chirag field. Amoco's local representative was hearing that BP had successfully recruited and mobilized former British prime minister Baroness Margaret Thatcher to make the case that BP should be awarded the rights to that field. They ultimately succeeded.

Meanwhile, several high-ranking BP executives abandoned the British company for Pennzoil. Following a similar script, but without the benefit of Baroness Thatcher's intervention, they convinced the Azerbaijan government to provide them the Guneshli field development rights. All three fields—Azeri, Chirag, and Guneshli (ACG)—were in close proximity, and Amoco's geoscientists believed that they might represent a single multibillion-barrel oil reservoir.

The stage was now set for a monumental struggle between Amoco, BP, and Pennzoil for supremacy, both for the order of field development and for access to the lone drilling rig in the Caspian Sea capable of safely exploring and developing the giant offshore fields. Alternative rigs would need to be disassembled and either flown in to be reconstructed or floated down the Volga–Don Canal to reach the landlocked Caspian. This process would take time and be very expensive; hence, access to the rig already in country was imperative in this competitive race.

Multiple companies vying for supremacy was a perfect scenario for the Azeris. They could sit back and watch the competitors go at one another "tooth and nail." They would not be disappointed. Each company tried to outdo the others by offering concessions and promising other benefits to the government in exchange for preferential treatment. SOCAR quickly realized that it could turn the competitive environment to its advantage. No demand from the government, it seemed, was too much for the companies to accept as each scrambled for the upper hand. Real economic interests were at stake, such as which field would be developed first, and that decision was of particular consequence given the availability in the short term of a solitary drilling rig.

The danger of this intercompany bickering to the FOCs, of course, was that all of the mounting concessions must be examined in the aggregate to determine what potential economic rent remained. If the end result was that the FOCs had no economic incentive to proceed with the project, then nobody would benefit, including the government. The obvious solution was "unitization."

In simplistic terms, unitization was the exercise of drawing a border around the three fields (which, as previously noted, might actually be a single reservoir anyway) and assigning percentage interests in the newly designed unit to each of the companies involved in the original fields.

Instead of having a larger interest in one of the constituent fields, each company would now be awarded a diminished percentage of the much larger unit. Unitization would allow the companies to construct a development plan for the unit as a whole, which made the most economic and technical sense, without regard to artificially and arbitrarily imposed field boundaries. It would also allow them to develop first those portions of the unit that offered the greatest economic value (the "sweet spots").

Unitization might sound reasonably straightforward, but it was anything but. It represented a three-way "zero sum game" struggle among the companies. Each firm presented its own self-serving scheme for unitization which, if accepted and implemented, would result in assigning the greatest petroleum reserves and production levels to it. Each company's recommended approach was bolstered by scientific evidence and expert testimony. The whole process must have been thoroughly mystifying and daunting to SOCAR, which, in turn, relied upon its advisors for guidance.

⊙ NEGOTIATIONS PRINCIPLES

Competition profoundly affects the negotiating process. It helps to ensure that the good or service being provided—in this case, the development of an offshore oil field—is priced efficiently in the marketplace. However, sometimes in their eagerness to win a competition, companies unwittingly concede all the remaining profit in the form of continual concessions. Some bids are priced so aggressively in order to win in a competitive setting that the victorious company struggles to convert its successful bid into a commercially viable project. Some companies have employed the questionable tactic of "getting a foot in the door" and then trying to revise their initial bid. This sort of "bait and switch" tactic can easily backfire, resulting in a substantial erosion of trust and credibility.

Another somewhat similar principle that comes to mind in regard to the Azerbaijan discussions is the importance of agreeing to a set of terms and conditions as a package rather than piecemeal and ad hoc. Each concession may seem fine on its own, but when they are considered as a package, they may again render a project uneconomic. It is important to convey to the other side that you are not committing to any terms until they are viewed as a package in the end. This "health warning" can help avoid subsequent misunderstandings and accusations that a negotiator is somehow "going back on his word."

Negotiations, as we have seen, are also profoundly affected by the relative power of the participants, which defines the negotiating leverage they may have. If but one company were negotiating with the state, the relative power would be better balanced between the two parties. The Azeris needed the capital resources, technological capability, and managerial expertise that the Western company would bring, and the energy company would need the approvals to explore for, develop, transport, and sell hydrocarbons that only the state could provide. In the case where the FOCs were competing so fiercely among themselves in their attempts to please the government, they neutralized some of the power they would otherwise have wielded.

Moreover, when a group of companies collectively exhibits conciliatory negotiating behavior by agreeing to make a string of concessions in the face of mounting competitive pressure and demands from the government energy company, it is easy to see what happens. It is akin to two dueling sword fighters going at it with a cliff behind the passive one. The passive fighter (representing the FOCs) keeps backing up as the more aggressive negotiator (SOCAR on behalf of the government) advances, continually pressing for advantage. The cliff, symbolizing a project that is no longer economically viable, looms ever closer. Potential profits get squeezed, and the distance to the edge of the cliff keeps diminishing. And so the story goes. The companies agreed that they needed to "draw a line in the sand," but none was prepared actually to take this step for fear of losing competitive advantage.

Sometimes the lead negotiator for a company underestimates her own power. It is tempting to assume that the government holds all the cards and that the company must do whatever it requests. In reality, that is usually not the case. In this instance, the government clearly needed the expertise and financial resources the companies could bring. If a negotiator assumes mistakenly that she is in a weak position, the result will likely be a suboptimal outcome that the company will later regret. This scenario can be exacerbated by the pressure she may feel from her own management to conclude a deal quickly. At some future point, company management will inevitably inquire, "Who negotiated this deal anyway?" In large corporations where transfers of managers to new roles occur every couple years, the negotiator will likely have moved on to a new assignment and may not be held directly accountable for the poor deal she put in place.

Many governments have learned that companies are frequently under time pressure to conclude a deal. Accordingly, they can employ a dilatory strategy designed to squeeze the negotiator and thereby extract additional concessions. Governments usually have a longer time horizon than corporate entities, which sometimes are too fixated on meeting near-term objectives that can be trumpeted to the financial markets for anticipated share price appreciation. The governments also know that despite a negotiator's stated willingness of his company to walk away from an investment opportunity, to do so may be viewed in the financial community as a failure and within the company as a "black mark" against the negotiator for being inflexible or unreasonable. In other words, it is easier sometimes for an individual in a negotiation to walk away than it would be for a company, especially if the opportunity is seen as a "company maker," as the Caspian project was for each of the FOCs in this case. In making a decision to walk away, one

> must examine what options exist and compare the attractiveness of proceeding with the project in question against the next best alternative. Again, in this case, the companies saw few opportunities globally that offered as much potential value. Perhaps that is one of the reasons that the parties worked so diligently to find a win/win outcome. Being directly involved in the negotiations, I know that it would have been exceedingly difficult to walk away. That said, the FOCs needed to show their growing frustration as a group, which was more easily accomplished post-unitization.

Ultimately, the companies realized that their collective behavior was self-defeating, and the government came to appreciate that continued delays threatened the project going forward at all. Abruptly one day in May 1993 the government published a list of interests in the unitized area, which appeared to be rather arbitrary and driven more by political considerations than by application of sound scientific principles.[3] However determined, the published interests in the unit were final. From that point on, the negotiations evolved into a bilateral affair—the FOCs versus SOCAR, which was empowered to negotiate on behalf of the state.

FIVE

Blind Alleys and Threats

COMPLETION OF THE UNITIZATION PROCESS WAS A MAJOR milestone that cleared the path forward to a comprehensive agreement between the two sides in the negotiation.

Well, almost . . . The negotiations were hijacked twice, leading the foreign oil companies down two unrelated blind alleys. In the first instance, which actually occurred prior to unitization, a Dutch oil trader named John Deuss and his company, Transworld Oil Ltd., injected themselves into the Azeri field negotiations with Amoco and its consortium. They were invited to the table by the Azerbaijan authorities, who were apparently displeased with the pace of the discussions and believed that Deuss could be useful in brokering the transaction. He had performed a similar role, advising the government of the Republic of Kazakhstan in negotiations with Chevron over the giant Tengiz sour gas field (meaning it contained high levels of hydrogen sulfide) located on the northeastern shore of the Caspian Sea.[1]

Over a period of several months, Deuss orchestrated a couple negotiating sessions, including one in Bermuda where he maintained one of his homes. That session began one Friday morning when my secretary pulled me out of a meeting. She indicated that John Deuss was on the telephone.

"Hello, Bill? John Deuss. I just wanted to check to see if you and your colleagues could come to Bermuda today to continue our negotiations with the Azeris over the weekend. The delegation from Baku is already here," the Dutch trader declared with his distinctive accent.

"I think so, but I will need to confirm with my team," I responded, though I suspected this would be one business trip where little persuasion would be required.

"Great! Just call me back to confirm and I will send my jet over to pick you up in Houston around noon."

Deuss's flamboyant penchant for "high living included stables with champion jumping horses, two Gulfstream jets, yachts, ski resorts, and a variety of homes."[2] With two jets at his disposal, he could realistically travel whenever or wherever in the world he desired.

After quickly conferring with my team, I discovered that the key people were indeed available, so I rushed home to pack.

Over the course of the weekend, Deuss alternated between separate discussions with the FOCs and with SOCAR in a sort of shuttle diplomacy, never allowing the two sides to meet except socially near the end of our stay. It was an interesting approach, but one that did not allow us to judge firsthand the reactions of the Azeris to our proposals. My assessment of Deuss himself was that he was bright and aggressive. He could at once be charming and gracious, yet equally he harbored a quick temper and a short attention span. It was a fascinating experience, though not terribly efficient, as the FOC team would spend lengthy intervals "cooling their heels" at their hotel while Deuss conferred with the Azeris.

The visit did include one memorable social occasion at which Deuss strictly prohibited business conversations. Both sides were invited to dinner aboard his 180-foot three-masted schooner, *Fleurtje*. This gorgeous tall ship was a testament to his considerable fortune. The Azeris were clearly awestruck by his display of wealth and power. However, the FOC representatives were not convinced that any of this would translate into a more rapid pace of negotiations or that his motives were in sync with SOCAR's or our goals. In fact, we remained suspicious of Deuss, who seemed to be trying to control the export pipeline options out of the Caspian, a move that ultimately could allow him to set tariffs at lofty levels. In December 1987, he reportedly tried to corner the Brent crude market, losing about US $200 million in the process.[3] Accordingly, Deuss's past behavior lent some credence to our current concerns.

Our exposure to the "rich and famous" concluded as quickly as it began. As the weekend wound down, Deuss's jet whisked us back to Houston (and reality!). As we feared, little progress had been made in the business discussions. And it got worse—the oil trader was sufficiently impressed with our lawyer, Michael, that he quickly snatched him away from Amoco, leaving us to bring another attorney up to speed. In the end, Deuss's involvement failed to act as a catalyst in the negotiations, and the Azeris soon removed him and Transworld Oil from the playing board. Direct face-to-face discussions between the FOCs and SOCAR resumed.

According to the Bermuda newspaper, *The Royal Gazette*, Deuss would later be arrested on the island nation in 2006 and extradited voluntarily to the Netherlands, his native country, where he stood accused of membership in a criminal organization, banking without a license, and failing to report unusual transactions. However, per the Dutch newspaper *De Telegraaf*, the Dutch courts found insufficient evidence to prove the allegations of his participation in a criminal organization. In May 2012, though, the Dutch trader and his sister, Tineke, were convicted of the two remaining charges.[4] They each received a six-month suspended sentence and were each fined 327,000 euros.[5] In August 2013, it was reported that Deuss and the Dutch government had agreed to a US $47 million settlement of a fourth charge that related to Deuss's involvement with the First Curaçao International Bank. The Dutch authorities alleged that he was engaged in activity that resembled money laundering and had participated in carousel fraud, a "scheme in which culprits falsely claim VAT payouts, and seek refunds from the government."[6]

The second time that the Azerbaijan negotiations were hijacked occurred post-unitization. In this instance, an Azeri businessman named Marat Manafov, who by then resided in Slovakia, turned up uninvited and unannounced at a presentation the FOCs were making in Baku, claiming that he had been appointed by the Azeri government to represent them in the negotiations. Tom Doss, who headed the Amoco Caspian project team at that point, and I agreed that we needed to corroborate the story directly with the Azeri government before dealing with Manafov and his colleagues. A call to then-Azerbaijan president Heydar Aliyev's office confirmed that he had indeed placed Manafov in charge of the negotiations on behalf of the republic. President Aliyev had reportedly reached this decision because he was displeased with the direction the negotiations were taking and with the proportion of the project value the proposed agreement would allocate to the Azerbaijan government.

As a result, the negotiations took place for a time in London and subsequently in the Slovakian resort city of Piešťany, where Manafov's company was headquartered. At one point, during the discussions in London, Manafov and his associates demanded what sounded decidedly like a bribe, insisting that the negotiations would "go far more smoothly and more quickly if the FOCs made a payment" in excess of US $300 million. The FOCs stood united in their absolute refusal to make such a payment, at which time the threats became more personal, directed against the negotiators and their families.

At an evening meeting in London that I attended, Manafov, apparently incensed that the FOCs had refused to meet his payment demand, stated that he was insulted and that he knew where all of the FOC negotiators and their families lived. I can recall personally fearing for the well-being of my family, and I called my wife Jan that night from London. She wanted to know if she and the children should go live for a time with her parents. I responded that I thought it was unlikely that Manafov would actually follow through with his threats but that I would keep her informed as events unfolded.

Manafov was unaware that an FOC representative in attendance at the meeting that night had secretly decided to record the proceedings with a device concealed in his pocket. It was a courageous step, given Manafov's unpredictable, erratic, irascible, and at times violent personality. The FOC representative was seated near the rear of the room. As a result, although the tape was turned over to the security department at one of the companies, the quality of the recording was too poor to make out its contents. Consequently, nothing came of this valiant effort to catch the Slovakian businessman in the act of soliciting a bribe and threatening the FOCs.

Manafov announced that future meetings would be held in Piešt'any. Given the events that had just transpired, we were apprehensive of trading the comparative safety of London for a Slovakian venue where Manafov would have far more control and influence. We also remained concerned that our internal discussions and communications might be intercepted and compromised. We began traveling with security personnel who were experienced in the detection and removal of electronic eavesdropping devices. Conference rooms would be swept routinely for such devices prior to all meetings.

Several colleagues reported incidents of Manafov brandishing a pistol, though thankfully that did not happen in my presence. One FOC participant in the Piešt'any discussions, who was picked up at an airport in Slovakia by one of Manafov's people, confided to me that when the vehicle's trunk was opened, he noticed that it was "full of weapons."

In the case of both blind alleys, the government discovered (ultimately) that the participation of third parties in the discussions actually undermined progress, particularly when the interlopers were primarily pursuing their own agendas that were not necessarily aligned with the interests of the Azerbaijan Republic. In the Manafov situation, it took a coordinated effort by the American, British, and Turkish ambassadors to convince the Azeris to abandon this approach. Turkey had considerable influence in Azerbaijan. Many of Azerbaijan's citizens were of Turkish ethnicity; the official language of the country was Turkic in origin; the Turkish company TPAO had a small interest in the unitized ACG project; and one of the routes for the proposed export pipeline traveled through Turkey. The US government was paying particularly close attention to the negotiations in the Caspian since it regarded the giant oil field as a means to diversify its sources of international crude oil and to reduce its reliance on the turbulent Middle East.

The agreement negotiated between Manafov and the FOCs, which they hoped would be ratified by the Azeri Parliament, was pronounced dead on arrival by SOCAR, and "Manafov was banished, apparently without compensation, in the wake of his failed agreement."[7] According to one report, he then disappeared in "uncertain circumstances in Bratislava, Slovakia."[8]

◉ NEGOTIATIONS PRINCIPLES

In the case of both of these "blind alleys," third parties convinced the government that they could better represent Azerbaijan's interest and obtain an enhanced outcome from the FOCs. Governments that lack experience with these sorts of negotiations are particularly susceptible to such contentions. However, it appeared in both cases that the third parties, not surprisingly, had their own interests at heart—ostensibly controlling the export pipeline from the Caspian area in Deuss's case and extracting an illegal bribe, in the case of Manafov. In both instances, the government ultimately realized that its interests were not being placed first and decided to return to direct discussions. In the case of Manafov, the intervention of the FOCs' home governments was a key tactic that helped produce the desired outcome more quickly. Sometimes governments are more willing to listen to their counterparts from other nations than to parties who have an economic stake in the outcome of the negotiations. The FOCs maintained a commitment to ethical negotiations throughout, an essential element in developing trust over the long term with the host government.

Negotiators frequently employ different styles of negotiations. Manafov, for example, decided to use overt threats. This approach backfired ultimately when the host government retracted his authority to negotiate on its behalf. Sometimes, however, implied threats, which are carefully worded, can be appropriate and effective. A government that feels a company is dragging its feet in negotiations for no good reason can show some irritation and remind a company that competitors are ready to take their place. A company that sees a government enacting arbitrary or punitive laws can advise the government that the risk-weighted economics of the project are approaching a point where its management may have to reassess its commitment to a project or not make incremental investments that would benefit both parties.

The shuttle diplomacy tactic employed by Deuss in Bermuda never allowed either side to hear directly from the other. The absence of direct face-to-face negotiations during this period meant that Deuss could put his own spin on reactions from both sides. There may be times this sort of negotiation is appropriate where the vitriol between two sides requires it. Discussions between the Israelis and the Palestinians come to mind as an example where shuttle diplomacy has been utilized from time to time.

Another point worth noting is that the impressive scale of an oil field means little if there is no available economic means to monetize the production. In Azerbaijan, there was a very limited market for the oil and few creditworthy purchasers, meaning that the production needed to be transported to distant markets. The cost of that transportation, in terms of the cost of pipeline construction and the associated tariff, is sometimes overlooked early in negotiations. The field, absent an export pipeline, would constitute a stranded asset.

SIX

Dangling the Lure of an Oil Export Pipeline through Georgia

ONE OF THE MOST CHALLENGING ASPECTS OF THE UNITized Azeri-Chirag-Guneshli project in the Azerbaijan sector of the Caspian Sea was to find an economic and secure means to transport the produced oil to Western markets. Given the magnitude of the potential reserves (perhaps in excess of five billion barrels), limited local demand, and the lack of creditworthy purchasers in the region, export was the only real way to monetize the crude oil in the near to medium term.

That said, there was no clear-cut reliable pipeline route out of Azerbaijan; each option had a major flaw. Iran lay to the south, but American companies were barred from participating in a pipeline across that country's territory due to US sanctions. To the southwest, Azerbaijan was still effectively at war with Armenia over Nagorno-Karabakh. To the north, an established pipeline hub existed in southern Russia at Grozny, but unrest was already smoldering there in advance of an all-out conflict between the Russians and Chechen separatists who were intent on establishing an independent homeland for themselves. A "mother of all pipelines" option across Turkmenistan, Kazakhstan, and on east to China was briefly considered but discarded as impractical—politically, technically, economically, and financially. Ostensibly, the only viable route out to Western markets lay due west across Georgia, either following the route selected by the Nobel brothers to Batumi on the eastern coast of the Black Sea, or traversing Georgia and Turkey to the Mediterranean port of Ceyhan. Ultimately, the latter routing was selected. Although it was longer and crossed some rugged mountainous terrain, it would avoid the tankering of crude through the narrow and environmentally sensitive Bosphorus Straits, the site of some beautiful coastline and expensive homes. The companies carried out extensive studies to determine the maximum size vessel that could pass through these narrow waters. They concluded that they needed fairly large tankers to allow for an economically efficient process, but the larger the vessel, the greater the chance of an incident (such as a spill) and the more severe the potential environmental impact.

The choice of Georgia was not without its own issues. That country would need to be convinced that a trunk line across its territory made sense and would benefit them. There was also growing ethnic violence in Georgia where the minority Ossetians, reportedly supported covertly by the Russians, were in a struggle with the Tbilisi-based Georgian government of President Eduard Shevardnadze. At the time, the president was perhaps best known internationally as the last foreign minister of the Soviet Union. Russia was not keen on the newly independent republics, such as Georgia, having independent sources of income. Russia far preferred the notion of a pipeline through its territory where it could extract tariff revenue and control the flow of oil, thereby exerting political pressure on the Azeris and Georgians.

Several colleagues and I arranged an initial trip to the Georgian capital of Tbilisi in 1991, long before unitization had occurred. Those were the days when Amoco was concerned only about exporting crude from its Azeri field. Georgia had no significant oil and gas production and therefore no real established energy industry. Still, the Georgians were well aware of developments in neighboring Azerbaijan. They knew that they needed to steer clear of the divisive Nagorno-Karabakh regional conflict. The newly independent republics would need to work together or risk being "reacquired," outright or effectively, by the Russian Bear to the north. To paraphrase and adapt an admonition attributed to Benjamin Franklin, "If the newly independent republics did not hang together, they would most certainly hang separately."

We discovered on our first visit that there was conceptual support for a pipeline through Georgia, but it quickly became clear that the only person who could make a decision on this weighty issue was President Shevardnadze himself. Accordingly, we scheduled a return trip, led by Mr. Blanton, with the specific goal of meeting with the Georgian president to promote the notion of such a pipeline.

The Georgian capital of Tbilisi was a lovely city. The streets were lined with sprawling shade trees and an abundance of attractive buildings. European influence was quite evident and pervasive. By the time of our second visit, however, parts of the city had been attacked and burned by Ossetian rebels, and there had been an unsuccessful assassination attempt on President Shevardnadze. His aides have long connected this and two subsequent failed attempts to Moscow's displeasure with his independent policies and sympathy with the cause of Chechen rebels.[1] Since President Shevardnadze's headquarters had been heavily damaged, he chose to ensconce himself in a well-fortified temporary facility about a mile away.

As we arrived for our meeting, we were met by a number of guards toting AK-47s. After we were thoroughly searched and patted down, we were allowed in and warmly greeted by the president. We found him well briefed and acutely aware of the vital importance such a pipeline project could play in ensuring his country's financial independence. He still possessed a wry sense of humor despite the recent attack on his headquarters. Far from being in opposition to an export pipeline traversing his country, he was, in fact, an ardent and vocal supporter.

After taking our seats at a long conference table in the president's office, we engaged in a thirty-minute discussion of the magnitude, complexity, and cost of the proposed pipeline project. We then explained that the first step would be to allow an Amoco team to visit the country to conduct a pipeline survey. We estimated that the process would take roughly a month.

"No," President Shevardnadze insisted, flashing his trademark smile. "It will take much longer than that."

Mr. Blanton, believing the president to be deadly serious, looked perplexed and slightly annoyed. "Why?" he asked.

"Your teams will need to stop in each village and sample our fine wines and Georgian hospitality. You could not possibly complete your task in such a short time!" Reassured, Mr. Blanton broke into a broad grin and continued with his sales pitch. We received a green light to proceed and departed after about a forty-five-minute meeting. We finally could be optimistic that Caspian oil could find its way out to markets in Europe and beyond.

Young with Georgian President Eduard Shevardnadze

We were feted that evening by our Georgian hosts, who introduced us to their wines (a vast improvement over our SOCAR-imposed liquid diet of pure vodka in Azerbaijan). We also learned that the Georgian approach to toasting was considerably more structured than in Baku with the formal appointment of a *tamada*, or toast leader, at all such celebratory occasions.

⊘ NEGOTIATIONS PRINCIPLES

Once again, it was clear that the Caspian oil development project, of which the export pipeline was an absolutely vital component, was set against a geopolitical backdrop. Russia was wary of the former Soviet republics in the Caucasus becoming too independent, thereby reducing the Kremlin's influence in the region. Georgia had been making overtures to the US and other Western countries for support. As noted above, the widely held belief was that the Russian intelligence service was behind the assassination attempt on President Shevardnadze. On the other side of the coin, Georgia knew it needed income to replace the goods and services it previously received from Moscow. Pipeline tariffs would fit that bill rather nicely.

From a negotiations standpoint, we were keenly aware that if President Shevardnadze felt that Georgia was the only route under consideration, the tariff might be so steep as to greatly diminish the economic attractiveness of the oil field development project. Hence, we went out of our way to say that Georgia was just one of the possible routes, and that Grozny in Chechnya remained under consideration (we knew that would be the last thing President Shevardnadze would want to hear). This approach kept the possibility of potential competition alive. In theory, we could still turn our backs on the Georgian pipeline route.

The relative timing of the Caspian field development on the one hand and the export pipeline negotiations on the other would also be important. If the upstream project (i.e., field development) progressed too far in advance of the pipeline, it could undermine our negotiating position with the Georgians and leave the oil temporarily stranded. It was going to be a complex task, but the crude was worthless to both countries just sitting in the ground. Clearly, security along the pipeline route would also be an ongoing concern, as it would be a tempting target for those unfriendly to the Georgian government.

Involvement of multilateral institutions, such as the European Bank for Reconstruction and Development (EBRD) and the International Finance Corporation (a sister organization to the World Bank), was identified early on as a mechanism to mitigate some of the political risk and bring some stability and protection to the combined project. Most countries do not wish to alienate the very financial institutions they might need to support their own infrastructure projects. These multilaterals had a deserved reputation of being rather slow to move, but their benefits were equally well recognized and valued, and they were a good fit for this extremely expensive and risky venture.

The 1,800-kilometer (1,100-mile) Baku-Tbilisi-Ceyhan (BTC) Pipeline was completed in 2005 at a cost of nearly US $4 billion and with a capacity of over a million barrels of oil per day.[2] Commissioning ceremonies were held at the Sangachal terminal south of Baku on May 25, 2005. Among those in attendance were Azerbaijan President Ilham Aliyev, Georgian President Saakashvili, Turkish President Ahmet Sezer, and President Nazarbayev of Kazakhstan.[3]

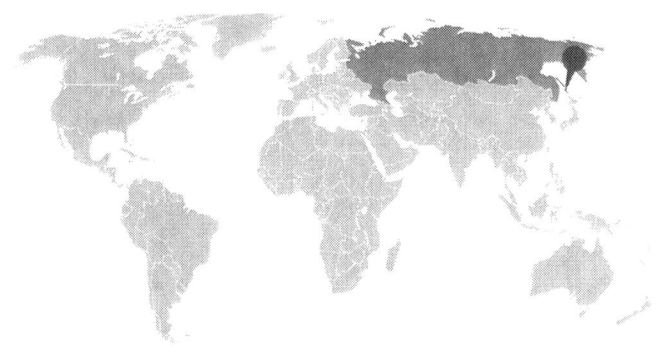

SEVEN

In Pursuit of Natural Gas off Russia's Sakhalin Island

MY FIRST TRIP TO THE RUSSIAN FAR EAST WAS A MEMORAble one. Two colleagues and I had been dispatched by Amoco management to gather information about a reported tender competition for offshore acreage to the east of Sakhalin Island and to purchase a data parcel—a prerequisite, as it was in Azerbaijan, to submitting a bid.

Sakhalin Island was the site of a number of sensitive Soviet military installations. It was also the location from which a heat-seeking missile was fired on September 1, 1983, by a Russian Su-15 fighter jet that brought down Korean Air Flight 007, killing all 269 passengers and crew on board.[1] The Russians claimed that the civilian airliner had strayed off course into Soviet airspace. Information from the flight data recorders, originally withheld by the Russians but eventually shared with investigators, suggested that the autopilot either failed or was not operating in the usual mode. In either case, the plane had indeed veered some 200 miles off course and had entered Soviet airspace.[2] Even after that tragic event, the Russians remained skittish about foreigners prowling around Sakhalin Island unchaperoned.

At the same time, an international dispute on the subject of the Kuril Islands continued to act as an irritant in relations between Russia and Japan. The Kuril Islands are a tectonically active chain of small islands that run from Hokkaido to Russia's Kamchatka Peninsula. During World War II, the Russians captured several of the southernmost Kuril Islands, which were Japanese territory, and have held on to them ever since. Accordingly, relations between the two nations continued to be strained, to say the least, and the Japanese government offered no "official" way for its citizens to visit the Russian Far East. However, the port of Niigata represented an unofficial "back door to Russia" and one of the most convenient ways for Americans to enter that part of the country via a flight to Khabarovsk. One must remember that Russia constitutes an immense land area. In fact, the distance from Moscow to Sakhalin Island (nearly 10,500 kilometers or 6,500 miles) is greater than that from Houston to Moscow (about 9,500 kilometers or 5,900 miles)! Consequently, the Russian Far East can most readily be accessed from the US by heading west.

In addition to the Japanese route, Alaskan Air operated a seasonal flight to Khabarovsk from Anchorage (via Magadan). As it was June, I chose to take advantage of this service rather than flying into Tokyo (Narita), taking a high-speed train to Niigata, and then boarding another flight to Khabarovsk.

I was quite excited about visiting Sakhalin Island and the Russian Far East for the first time. Alaskan Air typically flew over Nome, Alaska, and then across the 80-kilometer (50-mile) wide Bering Strait that separates the US from Russia. In the distant past, a land bridge connected the two continents at about this location. As is the case near the time of the summer solstice, the entire flight was accomplished in daylight or twilight. As I peered out the window of the Boeing 737, looking down at the virgin Russian landscape, I saw no evidence of human life below, not even a crude logging road or a frontier encampment.

As we approached Magadan, a few simple structures came into view.

Following our landing, we were greeted by the usual "follow me" truck that led our plane to its designated parking location (not that there was any shortage of parking spaces in this remote outpost). These follow-me vehicles facilitated communications when the ground staff might not be fluent in English, and the US crew was likely nonconversant in Russian. The plane eased to a stop, and the flight attendant prepared her bag of trash to hand to whomever was dispatched to greet us. I can still recall chuckling when she muttered aloud to a fellow flight attendant, "Where is the Jetway?" Not missing a beat, a passenger in the second row promptly responded in a slow Texas drawl, "Ma'am, this is Mag-a-DAN, not Dallas or Anchorage."

After a short layover, during which only a few passengers disembarked, we proceeded on to Khabarovsk, a commercially important urban center that, aside from the Pacific port city of Vladivostok, was the largest metropolis in this part of the USSR. The Russian Far East was one of the locations Stalin sent the disenfranchised during his reign of terror. Even though Khabarovsk lies near the border with China, virtually no trace of Asian heritage remains in the area. Those residents were presumably run off (or worse) years ago.

As if to dramatize the dearth of official communications between Japan and Russia, the time in Khabarovsk is one hour later than the time in Japan. That may not seem particularly noteworthy were it not for the fact that the Russian mainland lies due west of Japan, meaning the time there should be earlier, not later, than in Japan. Even more ironically, Sakhalin Island, which in turn lies to the east of Khabarovsk (and due north of the Japanese islands), is an hour later than Khabarovsk. In other words, when it is noon in Japan, it is 2:00 p.m. in Sakhalin, which lies in the same natural time zone as Tokyo! From the southernmost point of Sakhalin Island, it is possible on a clear day to make out the northern Japanese island of Hokkaido. Only the Cold War could create such a strange anomaly. To complicate time calculation even further, all airport clocks in the USSR were required to display Moscow time. When one was in East Asia, seven time zones removed from the Russian capital, having flight departures and arrivals stated in Moscow time was not terribly helpful.

Sakhalinmorneftegaz (SMNG), the Russian production association that managed exploration and production on Sakhalin Island and its surrounding waters, was to conduct a competitive tender for the Sakhalin II project similar to the process that had occurred in Azerbaijan. A bid committee, comprised of both local and Moscow-based members, would qualify potential bidders to compete and then evaluate all submitted bids. I met up with my colleagues (a geologist and a geophysicist) in Khabarovsk, since they had opted to access Khabarovsk via Japan, and we traveled together to Okha, where SMNG was based. Okha lies at the northernmost tip of Sakhalin Island. Its climate is affected a great deal by the presence of the Pacific and a cold ocean current. Even so, I was surprised to observe at this time of year the presence of icebergs lurking just offshore.

We traveled the last leg of our journey from Khabarovsk on an Aeroflot regional jet that had virtually no tread left on its tires. I wondered how well the poorly maintained aircraft would negotiate a slippery runway in Okha.

We met with Mr. Cherney (which means "black" in Russian), president of SMNG, and agreed without incident to purchase the data package. It was clear from our discussions with SMNG and its research institute that Sakhalin, perhaps owing to its great distance from the Russian capital, enjoyed a certain degree of autonomy and freedom from the shackles that Moscow usually imposed on those in closer proximity. Perhaps somewhat naively, we thought this separation from the Russian center might be useful. However, as we later learned, it was instead a source of considerable friction between Moscow and Okha. Bidders could unknowingly and unwittingly get caught in the crossfire.

Upon our return to Houston, management determined that we should form an international consortium to be more competitive, to exchange technical expertise and knowledge, and to share the immense costs of offshore exploration, development, and production in the inhospitable climatic conditions on the offered blocks (one- or two-year ice). Other companies, such as Marathon, McDermott, and Mitsui (known collectively as "3M"), were ahead of us in that regard and had been studying the area for some time. We reached out to BHP, the Australian company, and Hyundai, the South Korean concern, to form a joint bidding group. Our consortium became known as BAH (BHP–Amoco–Hyundai). We also believed that cultural diversity potentially might strengthen our appeal.

The Sakhalin project was particularly attractive because production would not need to be transported via pipeline across vast distances to where it would be consumed or exported. Production flowing through such pipelines can always be cut off arbitrarily for political or other reasons (as Ukraine later experienced), making projects in the Russian interior—even those lying on established pipeline routes—inherently more risky. Opportunities on the periphery of Russia somehow seemed a better bet. On the other hand, our geologists told us that the Sakhalin prospects were likely to be gas-prone. The lack of a natural gas pipeline grid meant that liquefied natural gas (LNG) would likely be the most efficient and economic method to monetize sizeable gas reserves. Once liquefied, the LNG could then be tankered around the world in specially designed vessels. Unfortunately, Amoco had yet to construct its first LNG plant (Atlantic LNG in the Caribbean nation of Trinidad and Tobago), and its partners likewise could offer little LNG expertise. This glaring weakness would prove problematic for the consortium.

Fresh off its experience in preparing a bid and presentation in Baku, Amoco worked with its joint venture partners to compile a winning bid. This process proved logistically complicated, as the teams from the three companies were so geographically dispersed—Melbourne, Seoul, and

Houston. As a consequence, meetings were infrequent and expensive. Moreover, we hailed from very different corporate cultures, which made finding common ground no small feat.

Needing to balance my ongoing negotiations in Baku with the preparations for the Sakhalin bid, I found myself working virtually 24/7. I made three around-the-world trips in the space of three months. During one telephone conversation with my boss, he memorably said to me, "Listen, while you are in Baku, could you possibly head to Melbourne for a consortium meeting when your sessions in Azerbaijan are in recess and then hit Sakhalin on the way home?" First of all, Baku is closer to Houston than it is to Melbourne, and Sakhalin is not exactly "on the way home" from Melbourne. Second, the airlines must have expected that nobody in his right mind would ever want to travel from Baku to Melbourne! I discovered that I had to backtrack from Baku all the way to London in order to proceed onward to Melbourne. After flying for two consecutive nights, I finally arrived in Melbourne at 5:00 a.m., just in time to shower, don a fresh set of clean but wrinkled clothes, and head off to an 8:00 a.m. consortium meeting. Much of the work in compiling the bid occurred in Australia.

We also spent long intervals on Sakhalin Island where we frequently encountered representatives of the firms with whom we were in direct competition (Exxon, Shell, and Marathon). During the time the BHP representatives were not working, they would amuse themselves by repeatedly watching their prized *Crocodile Dundee* video. There was not much to do on those long evenings on Sakhalin, and I was sure they were quite homesick.

During the winter months, when the Alaskan Air flight was unavailable, we would typically fly to Sakhalin from Niigata, Japan, via Khabarovsk. One return trip particularly stands out in my mind. A business colleague and I had boarded an Ilyushin-62 (Il-62) aircraft in Yuzhno-Sakhalinsk that was bound eventually for Moscow, though we expected to deplane in Khabarovsk. One of my colleagues told me that the Il-62 was the largest jetliner in the Soviet fleet when it was introduced in the early 1960s and could carry considerable cargo in its hold.

Passengers were standing in the aisles loading their carry-on items into the overhead bins when the plane abruptly roared down the runway and took off. Unlike the carefully choreographed procedure in the West, where all passengers must be seated with seat belts fastened before a plane can even push back from the Jetway, the Il-62 crew seemed oblivious to the fact that people were still stowing their gear.

As the plane banked sharply to the left, a few items fell out of the luggage bins, but nobody was seriously hurt. An unrestrained and frantic charcoal-gray terrier was running up and down the aisles at will. Apparently spooked by the plane's sudden takeoff and steep assent (typical for Soviet aircraft), it jumped into the lap of a well-dressed Japanese businessman. Before the gentleman could brush away the animal, it proceeded to empty its bladder on his hand-tailored business suit. He was not amused. The flight attendants did not make any move to assist, making one wonder if this sort of event was commonplace and part of the service on Aeroflot that customers had come to expect. A few fledgling offshoots from Aeroflot had begun to emerge, but realistically Russians had little choice but to rely upon the Soviet carrier for their domestic air travel at that time. For obvious reasons, I studiously avoided the carrier for long-haul trips, opting instead to enter the Soviet Union either from the east (to access Sakhalin) or from the west (to head to Moscow or Baku).

Following submission of its bid and its accompanying presentation, the consortium felt cautiously optimistic that it had a chance to best the heavily favored 3M consortium. The decision of the bid committee was inexplicably delayed, however, and it soon became apparent from media reports that the long-smoldering power struggle between Moscow and Sakhalin had burst into the open and now was on full public display. The internal Russian wrangling notwithstanding, press reports citing the usual "unnamed sources" indicated that 3M would be named as the winner "within days."

Then a uniquely Russian twist occurred in the plot. The bid evaluation committee itself was replaced by a new one, which applied different evaluation metrics and allegedly favored the BAH consortium bid. It was a shocking turn of events. As bizarre as it was, it was also short-lived, as the original bid committee, reinvigorated and spoiling for a fight, reasserted control, and 3M was summarily named the winner.

Amoco subsequently opened a local Sakhalin office in an attempt to latch onto the winning consortium. It was a shrewd move, but Amoco was stymied by its lack of LNG expertise. Shell, which had already become a world leader in LNG, employed the same tactic with far greater success. Ironically, under President Vladimir Putin, the Russian Federation would seek to drive out foreign investors such as Shell. Perhaps that is the Russian version of the "winner's curse." Sometimes one is indeed better off not winning a competition!

As runner-up in the Sakhalin tender, which meant little other than a source of pride, Amoco licked its wounds and refocused its attention once again on Azerbaijan. The end of the Sakhalin saga meant that I could also rededicate my efforts to the daunting task of negotiating all the agreements required for the Azeri venture.

⊙ NEGOTIATIONS PRINCIPLES

Joint ventures are rarely marriages made in heaven. Not surprisingly, companies normally focus first and foremost on their own corporate interests and objectives and only secondarily on the goals of the consortium. As long as these goals are well aligned, joint ventures can succeed, but frequently the alignment is more superficial than deep and real (or the parties' interests diverge with time), and, of course, cultural differences can further complicate matters. That was the case with the BAH consortium. Agreeing on all the elements of a joint bid was difficult enough; keeping participants in the presentation process "on message" was even more challenging.

It is tempting to allocate shareholding interests and voting equally among the parties, but then there is no clear leader who can make decisions and propel the consortium forward efficiently. Trying to lead such a team is akin to "herding cats" or managing a federation of nation-states such as the European Union that has no real decision-maker (Germany would be the closest) who can enforce fiscal discipline and control. Still, consortiums offer some advantages such as sharing costs and pooling technical expertise. They also reduce the number of potential competitors vying for the prize.

It is important to have a bidding agreement that spells out how decisions will be made in the period running up to a joint bid and how the scenario of one or more members dropping out prior to bid submission will be handled. Such an agreement inevitably must be concluded quite quickly, as the timing to submit a bid, it seems, is always shorter than the time available to accomplish the work. While a more permanent shareholder agreement will be compiled later, should the consortium win the competition, the precedents set in the bidding agreement frequently carry over into the more comprehensive and detailed shareholder contract. Thought also must be given as to how to terminate a joint venture equitably and efficiently if the interests do, in fact, diverge with time. Before entering any joint venture, a company should have a well-considered exit strategy that is fully within its control to execute.

Sometimes a company or consortium that does not win a competition can fashion a way to add itself to the winning group going forward. The 3M group won the competition, but Shell wangled its way into the team and effectively shared in the prize as if it had been there from Day One. Developing strategies for how to salvage victory from the jaws of defeat is incredibly important. Amoco nearly did the same thing but at the time lacked the key LNG experience that positioned Shell ideally to claim a share of the win. This was a huge lever for Shell since the project needed the LNG technology in order to transport the natural gas to overseas markets.

Finally, the Sakhalin experience was yet another reminder that bidding situations do not occur in a vacuum but rather against a political backdrop in the host country. The contestants could do little but witness from the sidelines the power struggle playing out between the competing bid committees. The underlying cause was probably the deep fissures that existed between the Russian center and local authorities. The companies quietly and unobtrusively tried to press for advantage vis-à-vis their competitors, all the while attempting to stay out of the crossfire.

A valuable lesson in Russia was that making friends with all factions was essential since one never knew which one would emerge with increased stature in the future. Make no enemies, and ensure that your bid contains elements that will appeal both to the Moscow and Sakhalin representatives on the bid committee.

During the pursuit of the Sakhalin project, and Amoco's unsuccessful gambit to attach itself to the winning consortium, I found myself making a number of trips to Yuzhno-Sakhalinsk, a reasonably attractive small city at the southern end of Sakhalin Island. The city had only one real hotel in which to stay that offered tolerable conditions—the Sakhalin Sapporo. The citizens of Yuzhno-Sakhalinsk reckoned that they would soon be rewarded with the opportunity to host the Winter Olympics. We did not want to pop their bubble by stating our belief that that outcome, while not totally implausible, was unlikely anytime soon.

We found little to do with our free time. We frequently dined at the hotel as there were few other worthwhile places to eat in town. Dinner was followed by an hour or two sitting in the hotel's karaoke bar enduring the experience of Japanese businessmen belting out familiar tunes. What they lacked in talent, they more than made up for with equal measures of volume, enthusiasm, and bravado. Let me just say that one has not lived until one has been exposed to karaoke participants at the Sakhalin Sapporo crooning the Beatles' hit "Let It Be." Never mind that the singing was incredibly off-key and risked shattering the windows of the establishment. Even if one chose not to enter the bar, the sounds reverberated and permeated throughout the hotel, making the option of retiring to bed early an unrealistic and unrealizable aspiration.

At the suggestion of the hotel staff, I took a short excursion out of town one day to watch the Russians fish for salmon. What posed as legitimate angling was actually a dump truck fitted with a crane. The "fisherman" would use the crane's bucket to scoop up water upstream from a weir. The crane would then pivot, releasing the catch into the rear of the dump truck. Not particularly sporting, but quite effective. The salmon caught on Sakhalin Island were delicious, as were the king crab. The Sakhalin waters, unlike those in Alaska, apparently had not been overfished thus far.

Many of my visits occurred during the winter months when Sakhalin received a tremendous amount of snowfall. One frigid and snowy night, two business colleagues and I were scheduled to dine with Mr. Cherney, president of SMNG. We were both courting the 3M consortium and SMNG during this time. Cherney had selected the restaurant and began the evening by ordering the requisite bottles of vodka. It turned out that my colleagues declined to drink that evening—one was a Mormon, who refrained for religious reasons, and the other demurred for reasons I cannot recall. In any case, I was left to shoulder the burden of consumption, further compounded by Cherney's pronouncement, "I do not trust anyone who does not drink."

Great! I thought. *We are here to garner his trust and respect and now it falls to me to shoulder the burden of downing the vodka.* Once again, I was left with the mission of "taking one for the team," never an enviable position. The only appetizers on the table were some sort of deep-fried smelt, which frankly tasted like it was not intended for human consumption but instead should have remained in the bait can. However, I needed something to coat my stomach and to absorb some of the alcohol, so I reluctantly partook.

The conversation seemed to go well enough. Cherney was in a good mood, prepared to indulge in my droning on about how Amoco's involvement in the Sakhalin II project would provide a substantial financial stimulus to the local economy. I felt more and more light-headed as the evening progressed. Although it seemed that the event would never end, mercifully, as with all business functions, it ultimately did.

I grabbed my coat and prepared to venture out into the subzero cold. I realized as I stepped out onto the icy metal landing that I was losing my footing, so I instinctively and reflexively grabbed on to the external stone wall (there was no handrail). It seemed like a grand idea at the time, but the building had been outfitted with Russian countermeasures to deter break-ins (read: shards of glass embedded into the wall). My lacerated hand began dripping blood almost immediately. I felt that I had but two viable options—either 1) put my hand in my pocket and ruin my business overcoat, or 2) keep it outside the garment where it would likely get frostbitten on the long walk back to the Sapporo Hotel. One of my colleagues offered a handkerchief, which I gratefully accepted, and wrapped it tightly around my hand. When we reached the hotel, about a mile's walk, we pooled the bandages we had in our medical kits and covered the wounds. I figured infection was probably unlikely since the laceration had bled so much.

Thinking back, I realized that my warm shoes had probably melted a thin veneer of ice on the landing, making it particularly slippery and treacherous. Of course, the vodka, which caused me to feel it was hot outside, notwithstanding the actual subzero temperatures, may well have been a contributing factor. I discovered over time that environmental, health, and safety issues were not of paramount concern in the former Soviet Union. In addition, safe and effective medical attention was difficult to find, particularly in a place like Sakhalin Island. We usually carried fairly comprehensive medical kits on our trips to Russia, including syringes, since the Russians had a nasty habit of reusing needles from patient to patient, which, of course, risked transmission of diseases like HIV/AIDS.

On this same visit, it snowed for the six consecutive days immediately preceding our departure, and we harbored real doubts that we would get off the island on schedule. Amoco had employed an Australian charter company to fly us in and out. A charter could fly a more direct route to and from Japan, and furthermore it avoided Amoco employees flying on Aeroflot. Increasingly, the Russian state airline was having trouble getting spare parts, and Amoco was becoming ever more uneasy about its staff relying on Aeroflot for transportation within Russia.

When we questioned the Aussie pilot about the prospects of getting off the ground, he responded, predictably and fully in character, "No worries, mite!"

When departure time came, the intensity of the snowfall had dramatically increased, and several inches of moisture-laden snow lay unplowed on the runway. The temperature was just below freezing as we piled aboard. Undaunted, our pilot went full throttle, and I confess I closed my eyes. We skidded a bit on the greasy runway but generally kept our forward trajectory. The visibility in the blinding snow had to be just a couple hundred feet. It was unlikely that the pilot could see any better or farther than we could. I also do not recall him communicating with the tower, but I doubt there were any other aircraft in the vicinity, certainly none crazy enough to take off or land in these conditions. As we climbed for what seemed like an eternity, eventually we emerged atop the cloud deck, and the sun, which had been missing in action for days, finally poked out. The pilot banked steeply to the right and steered our aircraft in a southerly direction toward the Japanese Islands, Tokyo's Narita International Airport, and the long journey back home to Houston.

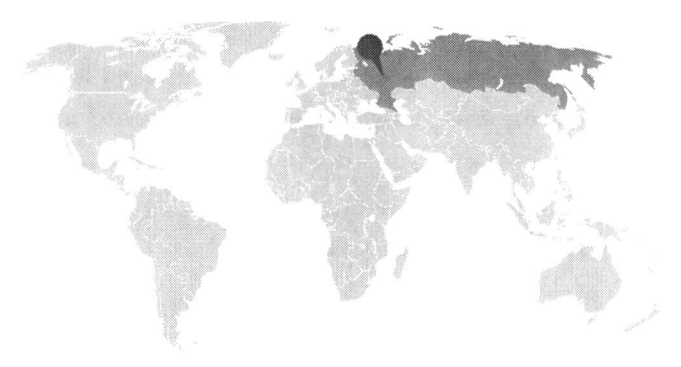

EIGHT

The "Wild West" Comes to Moscow

RUSSIA WAS NOT THE EASIEST OF DESTINATIONS IN THE early 1990s. Upon arrival at Sheremetyevo II Airport in Moscow, travelers were greeted first by mass chaos—arriving passengers pushing and shoving their way toward the passport control desks. There were no queues and, for that matter, really no organization at all. Rather, it was a test of maneuvering tactics and which passengers had the sharpest elbows. One might wait as long as two hours to reach the counter.

Then there was the ride to the hotel. Taking a cab from Sheremetyevo Airport was risky. A number of businessmen and -women had been taken to a remote area, robbed, and pushed out of their taxi. Companies had begun to send their own drivers to the airport to collect visitors and transport them to hotels in central Moscow. Some firms also leased dachas, comparatively posh and spacious single-family dwellings in wooded areas on the west side of town, to use as offices. These facilities, in many cases previously occupied by VIPs and party luminaries, offered enhanced security. The vast majority of visitors, though, still lodged in the city.

Like Baku in the early days, the Moscow hotels were uniformly dingy with filthy, threadbare carpeting and a unique and unpleasant odor. "Floor ladies" were stationed in the main hallway on each story of the hotel, presumably to act as the state's eyes and ears for what was going on and to pick up any useful intelligence. In one instance, I recall taking a wrong turn coming out of an elevator in Moscow only to have the floor lady inform me that my room was in the opposite direction. Clearly, she was paying attention to my comings and goings!

Muscovites now realized that to attract business, they needed first-rate lodging, and the facilities were modernizing rapidly. Following massive renovation, a number of hotels like the National reopened in resplendent form. Many, like the Kempinski and Metropol, to name a couple, contained first-rate restaurants. However, the downside was cost. Moscow had become one of the most expensive cities on earth, a cruel blow to the citizens of the capital, most of whom still earned meager incomes or were hapless pensioners. These hotels, and the restaurants within them, were effectively off-limits to the locals; the lucky ones landed low-paying jobs laboring in these facilities. It must have been more than a little distasteful to them to watch foreigners enjoying themselves in fine restaurants when they knew they could not afford them.

The ruble was another source of uncertainty and aggravation for Muscovites. It could—and in some cases did—lose considerable value overnight, leaving Russians even worse off, yet the opportunities to protect their savings by converting them into other currencies were largely restricted to the wealthy or well-connected.

At least the prevailing practice of maintaining separate entrances for hard and soft currency (i.e., the ruble) at Western restaurants like Pizza Hut had been eliminated. I always felt profoundly uncomfortable and guilt-ridden waiting in the much shorter hard currency queue, while those who were prepared to spend a couple weeks' wages for a taste of Western pizza stood in the cold, sometimes for hours. It would be perfectly understandable why the local population would resent privileged foreigners under these circumstances.

The main international hotel at first was the Mezhdunarodnaya

(International), located in the Armand Hammer Center (now World Trade Center). Dr. Hammer was a local hero, someone who early on recognized the potential of the Soviet Union and was prepared to make investments at the height of the Cold War. The "Mezh," as it was customarily abbreviated by foreigners, was certainly a marked improvement over the unrenovated dingy hotels like the Rossiya. In addition to hotel rooms, the Mezh contained offices, flats, and some retail space. Perched high above the Mezh's cavernous lobby, a glockenspiel played a tune several times a day while figures danced around it. Shoppers paused to look up at what must have been a real curiosity in Moscow at the time. In the hotel lobby was an Austrian café where I regularly dined. It served a reasonable pork schnitzel, a flavorful goulash soup, and to conclude the meal, a very tasty apple strudel. The café offered a reprive from the somewhat heavier Russian diet.

The Mezh did have its seedy side, however. Women desperate to earn some hard currency would frequent the Mezh nightly, hoping to seduce hotel patrons in order to support themselves and their children. Divorce rates were high in Russia, second only to the US in the world in the 1980s and increasing further in the 1990s.[1] Many single mothers were left to raise their children with little or no financial support from their former spouses. Many of these women were quite attractive and well educated, doctors and lawyers among them.

One of our lawyers related a story of how he received a knock at his door late one evening after he had already retired for the night. Upon opening the door, he was greeted by a stunningly beautiful twenty-year-old blonde, blue-eyed Russian in a short, tight, but tattered skirt who smiled and proclaimed, "I am dripping for you."

"Fine. Go drip somewhere else," he claims to have replied, though of course he produced no witnesses to corroborate his version of what actually happened when he found her standing in his threshold! If the story was as he portrayed it, the lady probably trudged off in disappointment but resolute to try a different line from her limited catalog of English phrases as she prepared to knock on yet another door. When I was an adolescent growing up in the United States, I never recall seeing such beautiful Russian women on TV stories about Moscow. I wondered if the American networks had exercised a degree of "self-censorship," showing only stocky, middle-aged Russian women.

Nightclubs had sprung up as well to provide entertainment for locals and foreigners alike. Some, like "Night Flight" on Tverskaya Street were joint ventures between Western and Russian entities, in this case Swedish/Russian. The women were largely Russian, while the male clientele were almost exclusively foreign businessmen (another opportunity for intelligence gathering by the state?). Other nightclubs were entirely indigenous businesses. Many were legitimate discos, while some were just fronts for prostitution. The practice of "face checks" began. Bouncers at the door would size up a potential patron to determine if he was likely to spend enough in the establishment to merit letting him in. It seemed that these nightclubs were off-limits to Russian males who presumably had little in the way of discretionary funds vis-à-vis their overseas counterparts anyway. There were no public smoking restrictions, so nightclubs were thick with secondhand cigarette smoke.

Complicating travel to Russia in the early days was the practice of issuing only single-entry visas. Instead of stamping the visa in a passport, the authorities would issue a stand-alone document, half of which was torn off upon entry and the remainder collected at departure. I surmised that this process was adopted to spare travelers any potential stigma or discomfort upon reentry to their home countries of having to explain why they visited the Soviet Union during the Cold War. Single-entry visas left no trace in the passport, save a couple staple holes. The problem with this approach, however, as I personally discovered, was that once a single-entry visa holder passed through immigration in the outbound direction, there was no returning.

On one occasion during the height of the Russian winter, I boarded a Lufthansa flight in Moscow, bound for Frankfurt, where I would change planes for my journey back to the US. It took quite some time to board the plane in Moscow and load all the baggage. In the meantime, it began snowing heavily and visibilities lowered dramatically. Not long thereafter, the announcement came from the cockpit, "*Meine Damen und Herren*, the Moscow aviation authority has closed the airport as a result of the inclement weather, and so our flight to Frankfurt will be delayed for some time."

I sat there wondering what would happen since most passengers, including me, had no visas to reenter the country. Sure enough, we were informed that we could not deplane because we lacked the required documentation and would have to ride out the snowstorm in the discomfort of our Airbus seats. Not surprisingly, we quickly consumed all the food and beverages on board; we then proceeded to watch the full allotment of films. Eight hours later, we eventually took off. Lufthansa was contrite and apologetic, but there was essentially nothing they could do. Multiple-entry visas soon became the norm, which avoided this sort of travel headache.

I always found it interesting that in Frankfurt for a time, following the dissolution of the Soviet Union, the authorities would run a Geiger counter over arriving passengers from Russia. Apparently, the Germans were fearful of travelers smuggling radioactive material out of the USSR. It was presumed that the previously tight controls over such substances in the Soviet Union had either been relaxed or eliminated entirely in practice. It seems unlikely, though, that would-be smugglers would risk placing radioactive isotopes in attaché cases without some sort of lead barrier. I guess anything is possible.

In Russia, passports were collected upon arrival at hotels and sometimes kept for several days. In one instance, in the eastern city of Khabarovsk, I managed to forget my passport, only realizing my omission once I arrived at the airport. Fortunately, I had allowed plenty of time to retrieve the passport from the hotel and still catch my flight.

Despite some inconveniences, Moscow remained a fascinating place to visit. Western symbols of wealth had begun to spring up, such as the Moscow Country Club. As if to prove that I was, in fact, a risk-taker, in contrast to my customary approach to life, I paid to go aloft in a hot air balloon. In reality, the risk was rather limited since the balloon remained tethered by a very long leash. Apparently, there had been prior incidents of zeppelins straying off course and alighting in forbidden locations like Red Square. Under the circumstances, I was more than happy to remain at the end of a long rope.

History was unfolding all around us. There was the dramatic 1993 attack on the Russian "White House," where the Duma (lower legislative house) met. There was also the opportunity to catch a glimpse of President Yeltsin's motorcade as it sped into the city from his west-side dacha, and to witness embryonic freedoms, such as speech and assembly, expanding ever so gradually. Most of these social advances were confined to the major cities of Moscow and St. Petersburg. Life in the countryside remained much as it had always been, and the peasants struggled mightily to make ends meet.

On one cab trip from Sheremetyevo II Airport to my hotel, the driver took me out of his way to view what had become of the statue of Felix Dzerzhinsky. Comrade Dzerzhinsky was a contemporary of Vladimir Lenin's among the early Bolsheviks. Lenin was well aware of Dzerzhinsky's revolutionary fervor, and trusted him implicitly and absolutely. Shortly after the Communists came to power in 1917, Lenin tasked Dzerzhinsky with setting up an internal secret police force to detect and crush any homegrown threats to the new Communist regime. Felix's response was to set up the Cheka, forerunner of the KGB, one of the world's most feared internal security forces and spying agencies.[2]

While the KGB fell out of favor during the Yeltsin era, it never was dismantled. It merely morphed into the Federal Security Service, or FSB. It was the same organization but with a new kinder, softer, gentler exterior—a secret police force for modern-day Russia. Its attention had begun to focus more on corporate espionage, but its priorities never strayed completely from state espionage or mischief-making on the world stage.

During the heady days following the short-lived Communist coup in 1991, Muscovites pulled down the statue of Felix Dzerzhinsky.[3] It had stood guard outside the dreaded Lubyanka prison in Moscow where so many Russian citizens had been imprisoned and tortured. Felix was a symbol of oppression and unbridled abuse of state power. The cabbie, brimming with pride as if he had personally played a role, pointed to the empty pedestal, promising not to charge me extra fare for the circuitous routing. He was good to his word.

As noted, some limited freedom of speech was creeping into Russian society, at least for those willing to take the risk of speaking out. At a conference in Vienna, I sat at a lunch table with a group of Russian journalists who were discussing their newfound freedoms. They exchanged jokes about the Russian government, while one lady journalist voiced her astonishment at how the environment had changed so unexpectedly and dramatically to allow such freedom of expression. During this era, smiles were far more prevalent than they had been. As I observed this new behavior, I asked myself if the process was irreversible. Could the genie be stuffed back in the bottle? Unfortunately, the answer during the Putin years was an unqualified "yes."

Russia's experiment with capitalism and free self-expression was not without its rough side, though. Moscow was a bit like a cross between the American Wild West and New York City during the height of organized crime. The so-called Russian Mafia sprung up to take advantage of the newfound liberties. Most of the violence was limited to power struggles among the Russian gangs, but as anywhere else, there was always the risk of getting caught in the crossfire.

One afternoon, several of my colleagues were having a drink at the bar in the Radisson Slavyanskaya Hotel near the Kiev train terminal when heavily armed police burst into the bar and ordered everyone down on the floor while they searched for suspected Mafia leaders. As you can imagine, my colleagues, unaware of the purpose or target of the raid, and lying flat on their stomachs, were quite frightened. No one dared to twist his head in order to look up to see who, if anyone, the security forces had apprehended.

It was not the only trouble at the Slavyanskaya. The American entrepreneur behind the development of one of the first Western-style hotels in Moscow, Paul Tatum, was defending his stake in the facility against "unscrupulous executives, the Chechen mafia, and a Russian business culture that wouldn't play fair," according to an AP report quoted in the *Los Angeles Times*. Late on the afternoon of November 3, 1996, Tatum was gunned down by a lone assailant carrying a Kalashnikov assault rifle concealed in a large plastic bag as Tatum entered the Moscow Metro located near the hotel entrance. He was hit eleven times in the back, hurtled down the steep underground steps, and died. "He forgot where he lived," Moscow police spokesman Yuri Tatarinov said later. "He tried to act in Moscow like he would act in the States or any other civilized country."[4]

Whenever I was in town, I regularly read the *Moscow Times*, a free English-language newspaper, and I remember being shocked to read a report about the shooting. On countless occasions, I had blithely trotted down those very same stairs to the Metro underground station on my way to a meeting or a restaurant. The Metro was an inexpensive and relatively safe means of transport around the sprawling metropolis. The lines and ornately decorated station stops were buried far below ground level in case Moscow was ever attacked.

Speaking of the Radisson Slavyanskaya Hotel, and on a distinctly lighter note, I would always enjoy getting a call from their business center. When the phone would ring in my hotel room, a soft, feminine, and sultry voice on the other end of the line would proclaim in a heavy Russian accent, "Comrade Young, we have a *fox* for you." Inevitably, it had the primal response of stimulating my senses and my imagination. I pictured a highly attractive blonde, with curves in all the right places, positioned provocatively in a lounge chair in the lobby. She was wearing a sleek, low-cut white dress and a single-strand pearl necklace. She smiled mischievously as I approached. But alas, reality returned with a vengeance. The call only concerned a *fax*—just a missive from the home office.

One Sunday, I was involved in an incident that highlighted to me just how dangerous Moscow had become. It was a delightful sunny day in July with just a few fair-weather clouds passing overhead. I had a craving for a Big Mac and so set out on foot for the nearest McDonald's restaurant on the far side of the Moscow River. It was about noon as I passed a bus stop on a wide thoroughfare clogged with traffic. Unbeknownst to me, several Gypsy teenagers had jumped off a crowded bus at the stop behind me and began trailing me, apparently attracted to the wallet bulging in my back pocket. Normally I wore a sport coat, which would have fully concealed the billfold, but as the weather was quite warm, I had dispensed with the coat.

By the time I realized that I was being followed too closely, the Gypsies jumped me and attempted to wrestle me to the ground. I fought back ferociously since I had no idea whether their objective was just to steal my belongings or to kill me. There were three of them, ranging in age from thirteen to sixteen, I would say. I landed a few good blows (it is amazing how strong one becomes when he feels his life may be in danger), but one attacker managed to rip the watch from my wrist. I began to spin, figuring it would be more difficult for them to get my wallet or to drag me down to the sidewalk. About that time, a Russian lady, hearing the commotion, leaned out the third-floor window of her apartment block, gestured with her hands, and shouted in Russian to the aggressors. They paused for a mere instant, allowing me to dart out into traffic on the busy street. I am not sure that I have ever run so fast, before or since. I felt like the fictional character Forrest Gump in the movie by the same name! I dodged the cars, continuing to peer apprehensively over my shoulder until I had reached the safe sanctuary of McDonald's. It was odd, I thought, but I had never considered the fast-food restaurant as a safe haven, a refuge in times of crisis!

I owed a debt of gratitude to that Russian lady, whoever she was and whatever she said. Physically I was unharmed, aside from a few bumps, bruises, and scrapes, but emotionally I was clearly shaken. I concluded that I needed to go out into the streets again right away to avoid becoming a prisoner of my own emotions and fears. That decision proved to be a wise one, and I recovered rather quickly from this episode. I sometimes chuckle when I reflect on the inexpensive travel watch the Gypsy gang captured. It was an exceedingly cheap one that I had purchased specifically with overseas trips in mind. My company cautioned its employees not to travel with valuable or ostentatious watches or other jewelry that might catch the attention of street thieves. At the time, gangs of primarily Gypsies were a problem in Moscow, especially around Red Square and other areas frequented by tourists.

After this incident, the company offered employees mace, but that policy was of no practical value to international negotiators who traveled on airlines where the substance was banned. Back home, the attack and response kept getting embellished by colleagues who wanted to rib me. As they recounted the tale—never mind that they were not present—I suddenly was keeping an entire mob at bay with my wild fisticuffs.

Muscovites were also in the early throes of experimentation with the laws of supply and demand and learning how to negotiate with those principles in mind. At Izmailovsky Park, on the outskirts of Moscow, a flea market operated each weekend on an immense open field. No matter the weather—even in the dead of winter—local artists and craftspeople would bring their wares to sell. There were Russian winter hats, called *ushankas* or *shapkas*, stitched together from the pelts of an extensive array of hapless furry animals, *matryoshka* dolls, oil paintings, and a wide assortment of other articles. Speaking of the *shapkas*, a Russian friend warned me that men should never wear them with the ear flaps down, no matter how brutally cold the weather might be. As he put it, "You do not want to look . . . how do you Americans say it? . . . like a wimp!" Not long after, I seem to recall seeing President Clinton on TV on a January 1994 trip to Moscow standing at an outdoor podium making remarks, his shapka flaps down over his ears. So much for quality pretrip protocol briefings!

The makers of the nested *matryoshka* dolls, in an effort to reach a larger consumer audience, had expanded their collections to include sets of NBA players (for example, Chicago Bulls Hall-of-Famer Michael Jordan) and the US presidents. It was fascinating how accurate the likenesses were of recent presidents, but as one opened each doll, following the line of succession back in time, the presidents began to bear an uncanny resemblance to the czars! Apparently, photos of the earlier American presidents were unavailable to the sculptors. With political correctness in mind, the smallest figure inside was a depiction of an American Indian, the first settlers on the North American continent. Also on offer were Christmas ornaments, statuettes of Father Christmas, and beautifully hand-painted lacquer boxes. Pieces of valuable antiques were also available, but many (especially religious items) were prohibited by Russian law from being exported. I could not help but wonder how they came to be on sale at the flea market in the first place.

I made the trek by the Moscow Metro to Izmailovsky Park on many occasions, including once when the daytime temperature was −23°C (−10°F). Shortly after the Soviet Union fell, the merchants were free to set the terms of sale for their goods, but they had little idea how to price them or how to go about negotiating. I can recall rescuing some beautiful landscape paintings from their perches in the snow, one for as little as

US $7. A frame back home easily cost ten times that much! Some artists negotiated; others stuck rigidly to their asking prices as they huddled near their small fires to keep warm. It was interesting to observe that some artists remained nonplused by the reality that their competitor had set up shop immediately adjacent to them and was selling essentially the same product for a much lower price or showed flexibility in the dickering process. I guessed they were content to await a purchaser who had not done his/her homework.

⊙ NEGOTIATIONS PRINCIPLES

To digress from the energy industry for a moment, if one has ever negotiated the purchase of a consumer good in a market or bazaar where bartering is allowed, and in some cases even expected, one quickly learns how the game is played. Each side puts forward a starting price, akin to placing goalposts on an American football field. The pair of initial prices defines the playing field (boundaries) where the negotiations will actually take place. Each side then offers proposals and counterproposals in an effort to close the gap and reach a mutually agreed price. Prior to commencing the negotiations, it is essential to have determined your bottom-line price so that you do not get dragged too far toward your own end zone, to continue the football analogy.

After a reasonable amount of dickering has occurred, and it is apparent that further flexibility by the seller is unlikely, it may be appropriate to say, "I will pay US $10 for this item but not a penny more." If there is no acceptance of this price, it is usually a good tactic in most cultures for you, quite literally, to begin to walk away. This action signifies that you feel that the gap is too large to be bridged. You are prepared, therefore, not to consummate the deal at a price higher than your last offer. If the price you offered to pay is not unreasonably low, the merchant will frequently chase off after you. If not, the final offer may well have been too low, leaving no margin for the seller, or for some other reason the seller was not sufficiently motivated to complete the transaction.

Of course, if there are many prospective purchasers prepared to step forward and pay the full asking price, your position may be substantially undermined. Take the case of a consumer trying to negotiate the purchase of a new Toyota Camry at a significant discount to the Manufacturer's Suggested Retail Price. I was in precisely that position a number of years ago. The Camry was in its second model year. Demand was high and the supply, all manufactured at the time in Japan, was quite limited. I recall the sales representative somewhat arrogantly pointing out the obvious, "Why would I sell a Camry to you at a reduced price? I have plenty of customers who are prepared to pay the full retail price that we are asking." Clearly, my leverage was somewhere between minimal and nonexistent. Luckily, I located another dealership that seemed to have less business and a considerably larger inventory of vehicles, and I was able to obtain a vehicle at a somewhat lower price.

The bargaining for souvenirs at Izmailovsky Park offers an excellent opportunity to mention some additional basic negotiations techniques. When the merchant initially divulges his asking price for an item, a "flinch" from the would-be buyer can be an important signal that the original asking price was totally unrealistic. It is important to appreciate that the seller's starting price frequently bears little relationship to its true market value. In fact, if anything, it is likely a multiple of what the article is actually worth.

It is a good idea to ask colleagues who have purchased souvenirs in the flea market previously what sort of multiple is commonly being employed. This intelligence is essential to formulating a reasonable and informed counteroffer. Should it be 50 percent of the initial number, 20 percent, 10 percent? The counteroffer should be well below the maximum the buyer would be prepared to pay.

It is amazing how many unwitting buyers take the original offer,

> discount it by some arbitrary percentage, and then split the difference to get to the final negotiated figure. Since sophisticated sellers are usually aware that would-be buyers use this technique, they set their starting price accordingly. This approach to negotiating almost certainly will result in the buyer paying too much. Another pitfall is for a buyer to jump too quickly to his bottom-line price.
>
> At the end of the day, if the merchant is unwilling to negotiate an acceptable price, remember that there are likely other merchants offering similar wares nearby. In any case, few souvenirs are indispensable anyway!

There was a scare at Izmailovsky Park on one occasion when Russian authorities discovered that radioactive material had been shallowly buried at the park. The authorities claimed the incident was the work of Chechen separatists, but one never knew whether that was the case or just an opportunity for the authorities in Moscow to place the blame where they wished.

With liberalization in Russia in the early 1990s, the open practice of religion was once again allowed, and the Muscovites set about reconstruction of the largest Russian Orthodox church in the world, the Cathedral of Christ the Savior, on the banks of the Moskva River. It was originally erected to celebrate Napoleon's retreat from the Soviet Union (success on the battlefield, the Russians believed, was the result of divine intervention). However, Stalin had the church dynamited in 1931 to make way for the Palace of the Soviets[5] and to make it abundantly clear, I suspect, that atheism was the state's only tolerated "religion." Reconstruction began in 1990, and the structure now stands as a gleaming testament to the resilience and resurgence of Christianity in Russia.

Sometimes Russian delegations would visit my company in the United States. When they did, they were usually paid a small Law

Department–approved per diem allowance to enable them to pay for incidentals while they were away from their homeland. They usually arrived virtually penniless, and Russian rubles could not be exchanged readily for US dollars anyway.

On one occasion, I took a couple of Russians to a popular Houston restaurant that featured various game selections like pheasant, quail, and rabbit. One of the Russians began to sob quietly but uncontrollably. When I asked what was wrong, the gray-haired seventy-year-old replied, "I have wasted all of my life in a bankrupt system. What do I have to show for it?" I struggled to come up with a suitable response since I fundamentally agreed with his lament.

———

Sometimes, one can become too experienced in a particular field for one's own good. Following Amoco's unsuccessful effort to win a stake in the Russian Sakhalin project, I decided that it was important to move on with my career by gaining experience in other areas. By that time, the Azeri negotiations seemed to be moving in the right direction and it was a good time to pass the baton. Amoco's Chicago-based Treasury Department contacted me informally to advise me that they would be offering me a position in corporate finance, subject only to Amoco Eurasia granting its permission to approach me. Finance was my area of academic concentration at the Wharton School at the University of Pennsylvania where I had earned my MBA. The proposed position, based in the Windy City, would enable me to refresh and revalidate my financial credentials. I reasoned that I would become pigeonholed if I continued to work solely on former Soviet Union project negotiations. The Chicago transfer would also include a promotion, which, of course, was of interest.

To my surprise, I was called into my boss's office and told that I was "indispensable" to the Azeri negotiations and hence my transfer and promotion were being blocked. I thought, *So this is how you reward someone who is essential to the negotiations process—by penalizing him?* I did not like the idea of being shackled to a project whose negotiations, despite having made big strides, could possibly drag on for years. Undaunted by the first rebuff, I managed to move to another Houston-based group (International Acquisitions and Divestitures) that provided the breadth of experience I was seeking.

About a year later, the Russia team contacted me, indicating that they needed an experienced international negotiator for another development venture, this one in Western Siberia. They inquired if I would be willing to return. Some added incentives were offered, so I somewhat reluctantly agreed. About eighteen months later, I would succeed in taking the Chicago-based treasury position.

⊘ NEGOTIATIONS PRINCIPLES

Negotiations may occur in your business dealings and in your private life, but they also are key to your career progression. In this case, I concluded that I was being increasingly viewed as an expert in a fairly narrow field (negotiations in the former Soviet Union, specifically Azerbaijan). When I sought the opportunity to expand my experience, which ultimately would benefit my employer in the future, I was greeted with a roadblock to my promotion and career progression, which I felt was a short-sighted response. I was seemingly a victim of my own success. I realized that my services, however, would be marketable beyond Amoco. I took the calculation that if I applied a second time for another position, the company would not risk blocking me again. It was brinksmanship once again with the implied threat that I would leave to take a position with another company if my initiative was thwarted again. My gamble paid off. I was able to accept the acquisitions position and then a year later mend fences with Amoco Eurasia by returning to lead a new negotiation.

Employees should not shy away from frank and open conversations with their bosses about career progression. Avoiding such discussions not only risks slower upward movement and potentially a lower compensation level, but it also sends a signal that the employee lacks confidence and the drive to get ahead. Of course, such conversations must be accompanied by a solid record of outstanding achievement and, where a promotion is sought, the skills and requisite experience to perform well at the next level.

Similarly, I have noticed that many people are reluctant to negotiate a starting salary with a prospective employer. So long as the applicant is respectful, professional, and reasonable in submitting a counterproposal to the original offer, the worst that can happen normally is a polite but firm "no." In many cases, the manager who extended the original offer has some flexibility and may even anticipate and expect a negotiation. Of course, in a hiring environment where there are far more applicants than slots to fill, the odds of success may be considerably lower. In such case, the applicant must assess the risk, however remote, that the company might withdraw its offer.

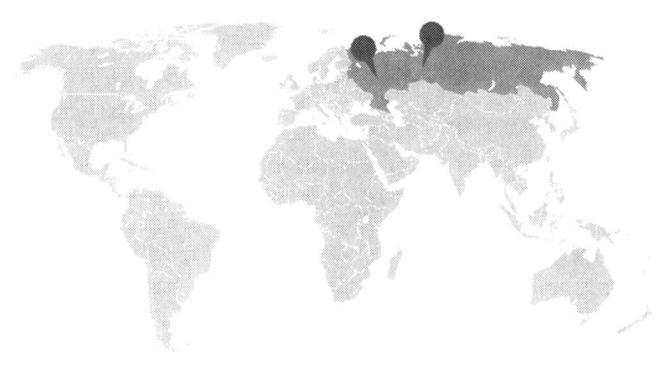

NINE

Oil or Nothing
Opportunity Knocks East of the Urals

IT APPEARED THAT I WAS DESTINED TO REPRISE MY ROLE AS an international negotiator, focusing primarily on Russian development ventures. The new project was called Priobskoye (pronounced: Pre-OBE-sky-ya) and comprised a discovered, but largely undeveloped, mega oil field in the Ob River flood plain, which was situated about 1,900 kilometers (1,200 miles) east–northeast of Moscow in Western Siberia. The field covered an area of over 3,200 square kilometers (2,000 square miles). This part of Russia, east of the Ural Mountains, was part of Asia.

Our coventurer in the project was the local production association, Yuganskneftegaz (YNG), which means "Yugansk Oil and Gas" in Russian. Most of the negotiations were to be held in Moscow, which, as we have said, was becoming an increasingly civilized place to do business, albeit extraordinarily expensive and still somewhat risky from a security standpoint. Boris Yeltsin had come to power, and the experiment with "Russian-style democracy" was in full swing. The Russians had no recent experience with the freedoms of assembly, speech, and press, and it was fun watching them test the new boundaries.

The Priobskoye negotiations process had two stages—first, reach agreement with partner YNG and second, negotiate together with the central and provincial (okrug) governments to obtain approval of the PSC. The first stage involved negotiating with YNG's director of new field developments, Vladimir Paltsev, a Russian military veteran and war hero from Russia's long, costly, and largely unsuccessful campaign against Afghanistan in the late 1970s and 1980s. Paltsev was a tough negotiator but a reasonable businessman who understood that Russia was changing rapidly and that partnering with a Western firm with the expertise and funds necessary to develop the Priobskoye field was in YNG's interests. A plethora of intense negotiating sessions took place in a large conference room situated at the rear of the Radisson Slavyanskaya Hotel. Amoco briefly maintained an office in the hotel before moving to more spacious quarters in a dacha on the west side of Moscow.

In addition to me as chief negotiator, the Amoco team typically consisted of a lawyer, a petroleum engineer/production specialist, a tax attorney, an economist, an accountant, and an interpreter. I would do much of the speaking on behalf of Amoco, but I would also schedule segments for the various experts to address portions of the draft agreements that fell into their disciplines, all the while trying to keep us on message. Many times we would be battling fatigue—jet lag is not easy to shake at any age—and hunger. To make matters worse, I would sleep poorly on the road, not a great situation for a negotiator who needed to be at the top of his game. My insomnia was not helped by the hotel's temperature being too high and the windows in the rooms not opening. In addition, the comforters on the beds, which worked effectively to trap heat, cooked me slowly overnight like a pot roast in a Crock-Pot. Sleep deprived, I was legendary for the number of caffeinated soft drinks I would consume during a lengthy day of discussions.

Lunches were the only real break during these sessions and were taken in a restaurant in the Slavyanskaya Hotel that offered a daily buffet.

With time, I knew exactly which items would be sitting where on the tables! In addition, Paltsev would call time out periodically to go outside for a smoke. Many Russians were notoriously heavy smokers, and I witnessed at least one international tobacco company on the streets of Moscow shamefully handing out free cigarettes to every passerby, regardless of age. The Amoco team would normally leave the premises for dinner, just for a change of scenery. Dining out during the winter months, of course, meant braving the bone-chilling cold of Moscow nights.

I got to know Paltsev rather well with time, and we developed an exceedingly good rapport. We even exchanged jokes about lawyers (which I gathered must be a universally accepted form of humor). In point of fact, humor is rather risky to employ across different cultures, unless a solid bond of friendship has been forged first. There is a distinct danger that a joke will not translate well or worse that the recipient may be inadvertently offended. On the rare occasions when I tried to inject some humor, I would have to wait for the translation into Russian before I knew whether or not my counterparts shared my sense of humor.

Unlike the Azeri negotiations, YNG recognized early on that it needed quality advice and had retained a Moscow-based consulting firm, CentreInvest, to do so. The representatives from the firm were in their thirties, well-educated, quick learners, and more familiar with Western economic concepts than many of YNG's personnel. It helped us immensely to be speaking the same "language" with respect to contractual and economic concepts.

We relied upon local contract interpreters supplied by an agency in Moscow since it was tough to justify the cost of sending over additional personnel from Houston. The drawback to this approach, of course, was that we were entrusting sensitive material with these interpreters who could well be working for the other side (in this case YNG) or the Russian KGB/FSB. Our Moscow office started hiring its own interpreters, but that did little to diminish the risk since these employees were drawn from the same labor pool. Ultimately, we did start flying Houston-based employees, who were fluent in Russian, to Moscow to function as interpreters. As noted earlier, Russian KGB operatives had by this time switched their primary focus from governmental spying to corporate espionage. The Russians were short of funds, and knowing the other side's position in a negotiation could come in very handy! Interestingly, many of the oil industry applicants had KGB experience on their resumes. They did not even attempt to hide it!

⊙ NEGOTIATIONS PRINCIPLES

In the game of contract bridge, there is an old saying that a "peek is worth two finesses." The adage applies equally well to negotiations. If the other side has perfect information as to your bottom-line position, no amount of bluffing or compromise is going to help. No sophisticated espionage techniques or equipment are required to get the upper hand. People have a tendency to leave papers sitting on the table, or they believe incorrectly that nobody can read English-language material that is upside down. (Of course, numbers require no translation!) Conversations can be overheard through thin walls in a breakout session or at a nearby table at lunch. Finally, there is the obvious method of an interpreter who has been present during internal meetings debriefing the other side. We were careful to lock up key documents when we would break for lunch, and we routinely excluded the interpreters from our internal deliberations, no matter how long we had been working with them and trusted their confidentiality and ethics. Hotel rooms and even trash bins are other locations of vulnerability. Written materials should be shredded, and sensitive data files should be password protected. For our most delicate discussions, we took long walks outdoors to diminish the risk that our communications might be intercepted.

Concern about sensitive information works both ways, of course, and the rather xenophobic Russian government was absolutely paranoid about their geologic maps. Maps were considered "state secrets," and it took approval from the upper echelon of the Russian government to share them with foreign companies that, in their view, could be spying for their home governments. Accordingly, we would be shown the results of seismic surveys and wells but would not be given the precise geographic coordinates to enable us to determine where on the planet they were located! This approach complicated life for our geologists and geophysicists.

The negotiator must maintain discipline at the table so that the team is speaking with one consistent voice. However, it is important as well to have the experts around the table address those portions of the agreement that fall in their disciplines or to catch the negotiator when he begins to misspeak. We had a standard procedure that any team member could call "time out" if he/she felt we were heading down a dangerous, incorrect, or unproductive path in our discussions or if we just needed to plot our next move. Such breaks were imperative. The discipline at the table was akin to choreographing a musical. One wanted to achieve a desired effect with all the moving pieces functioning in perfect harmony.

It is also important to have additional "eyes and ears" in any meeting. They function as witnesses and can also pick up subtleties coming from the other side of the table, which may be missed by a negotiator who is speaking much of the time. These additional participants are also in a good position to judge if the arguments being advanced from our side of the table are understood and resonating with the personnel across the table. Many times these witnesses caught misunderstandings early on before they could snowball. I was always a big proponent of the team approach to negotiations. Invariably, the group would fashion a strategy that was better than any one person would have, and I was grateful for their invaluable contributions.

> Whenever possible, it is helpful to avoid one-on-one negotiations, since these sessions have no witnesses to what has been stated or perhaps verbally agreed. It is always challenging to be thinking, speaking, and taking accurate notes simultaneously. The third item can be delegated to a team member so that the lead negotiator can focus on the other two activities.

The primary thrust of this first phase of negotiations was to reach agreement with YNG on a joint operating agreement (JOA) as well as on the text of a proposed PSC that could then be presented jointly to the Russian government. We have briefly touched on a PSC. A JOA provides the terms and conditions that govern the working relationship between or among the nongovernmental parties to the PSC as well as the rights and obligations of each party. Sometimes a single company is appointed operator; on other occasions the agreement may provide for a joint operation between or among two or more companies. On some occasions, a state oil and gas company is a participant in the JOA as well. If such is the case, then sometimes the nongovernment entities will form a separate group to discuss and agree in advance how they will vote at the operating committee meetings.

In this case, the YNG and Amoco negotiating teams had an excellent working relationship, and we succeeded in completing work on the JOA without much delay. We developed trust in each other, which, of course, was absolutely essential. In addition, YNG and Amoco shared a common objective—starting discussions with the government about the PSC as soon as practicable.

In a sense, we had begun to outrun the legal process in Russia. While the Duma (state legislature) had enacted a Petroleum Code, discussions to adopt implementing regulations had bogged down. Furthermore, the government had begun to fear that PSCs gave the foreign companies too great a share of the project net proceeds and too much power. The Russians wished to retain as much control as possible. We expended considerable time and effort trying to convince them that the proposed agreement was similar to others used commonly around the world and that the rights of the various parties were fully described and well protected. This tension, which exists in most host government/oil company negotiations, was heightened in Russia. Unfortunately, the combined effect of the foregoing factors was to slow considerably our progress in securing the requisite governmental approvals for our agreement.

We lobbied aggressively through our Washington office to get the Priobskoye project added to the Gore/Chernomyrdin list. At the time, Russian Prime Minister Viktor Chernomyrdin was US Vice President Al Gore's counterpart in the Kremlin. The list contained those investment projects in key sectors of the Russian economy that the two leaders agreed were high priorities for both governments. Messrs. Gore and Chernomyrdin further pledged to identify and remove any roadblocks that might arise.

I prepared an abstract of the Priobskoye project, together with our assessment of its strategic importance to both countries. Amoco then submitted it to the US Department of Commerce. From there, the document gradually snaked its way through the US government to the commerce secretary and ultimately to Vice President Gore's desk. YNG was working the same issue from the Russian side. We eventually succeeded in gaining both US and Russian endorsement of the Priobskoye project, and it was added to the list. This vital step helped the project gain some added impetus and traction in the lengthy approval process.

In April 1995, Paltsev and I made a very successful joint presentation to the Special Government Commission (SGC), which had been appointed by Prime Minister Chernomyrdin to consider the Priobskoye project. It took place at the Metropol Hotel in Moscow. Vadim Dvurechensky, Russia's deputy minister of fuel and energy, chaired the SGC. A well-respected oil and gas expert, Dvurechensky was also intimately involved in the process of drafting legislation to govern PSCs in Russia.

To celebrate our success, the president of Yuganskneftegaz, Mr. Ivanov, invited Paltsev and the Amoco team members to dinner that evening at the Mexican restaurant Santa Fe, a Venezuelan/Russian joint venture. Though the Mexican food was rather mediocre (the tortillas, for example, had a tendency to disintegrate when rolled), a meal at the Santa Fe was a great break from the daily Russian fare.

For the occasion, the Russians shed their insistence upon vodka, opting instead for tequila and margaritas. About halfway through the meal, Ivanov asked me if I liked to hunt. Recalling what had happened to Mr. Blanton in our earlier negotiations in Baku, I hesitated before responding. As I began to speak, our host interrupted, saying, "Yes, William, we must go *bear* hunting." I am not a hunter, and even if I were so inclined, I probably would not break into the sport by stalking those oversized and powerful animals!

Observing my apparent reluctance and/or fear, Ivanov stated rather matter-of-factly, "Not to worry! . . . We chain bear to *tree*!" *Not terribly sporting*, I thought. "Oh, good," I recall responding, sounding distinctly relieved, I am sure, but still unenthused. Fortunately, the bear hunting offer was never repeated in future encounters, probably having slipped from memory over an evening of heavy drinking. I certainly did not bring it up.

This celebratory occasion, like many business dinners, was long on food, talk, and alcohol. In this case, it culminated in Americans and Russians raucously downing shots of tequila. I was seated next to Gary, one of our drilling experts. He seemed to be leading the informal drinking contest, at least among the Americans, though none of us was in the same league as our Russian friends. As we left the dinner, Gary, with decidedly slurred speech, insisted upon negotiating the fare for the taxi ride back to our hotel.

"Mr. Bill, you always do all da negotiatin'. Let me show ya how a driller does it!" he stated proudly in his Texas drawl. Appreciating that his Russian was far more fluent than mine, even when he was inebriated, I readily acquiesced.

"Sure. Go for it, Gary!"

Wobbling noticeably as he approached the curb, he successfully hailed a cab, most of which were just private motorists with unlabeled cars moonlighting to make a little hard currency. The cabbie rolled down his window on the passenger side. Gary leaned over, peering into the vehicle, and asked in perfect Russian, "How much to the Mezhdunarodnaya Hotel?"

"Fifteen rubles," came the response from the dark interior of the Russian Lada.

"Twenty!" Gary insisted. I will never forget the inquisitive and bemused expression that crept across the cabbie's face. He probably was just beginning to grasp some of the basic tenets of capitalism and how to negotiate in a free-market world. He was clearly baffled by Gary's counteroffer, questioning his own understanding of the negotiations process. Never mind that we were not staying in the Mezhdunarodnaya Hotel anyway, but rather at the Slavyanksaya. After a pregnant pause as the cabbie pondered his next move, wondering if he had started the bidding too low, he acquiesced to the higher fare, and we clambered into the car. I told Gary to consider the extra five rubles a tip.

⊘ NEGOTIATIONS PRINCIPLES

It may be stating the obvious to suggest that just as drinking and driving do not mix, neither do drinking (or for that matter, any other impairment) and negotiating! In American parlance, if negotiations are about setting the goalposts to determine the playing field for the two sparring parties, Gary was sacked in his own end zone for a safety. Two points for the opposing team!

When negotiations occur between parties who do not share a

> common first language, it is important for each side to negotiate in its native tongue. It is far better to let well-qualified interpreters who have basic familiarity with the subject matter bear the burden of translating what you say.
>
> Sometimes inexperienced interpreters have a tendency to translate the English to Russian (or vice versa) too literally, with the actual intent being sacrificed in the process. Interpreters who have a commercial and/or legal background can help avoid this problem. Also, as mentioned earlier, it is imperative to take precautions so that one can be reasonably certain that the interpreter is not working for, and will not pass sensitive information to, the other side. Finally, whenever possible, a negotiator should insist that the official version of a contract be in his native language. Of course, sometimes the host government requires that the version of the contract in their language prevail in the event of a dispute. In such case, the next best alternative is to have both language versions of the agreement carry equal weight. Any differences that arise and that cannot be resolved amicably among the parties would be settled via the dispute resolution provisions of the agreement.

For the Priobskoye project to succeed, we appreciated that we would need support both at the federal and okrug (provincial) levels. To date, we had focused most of our attention on educating officials in Moscow, but we had spent comparatively little time in the local communities explaining how the project could benefit them. The company decided that it might be a good idea to sponsor a social reception in Khanty-Mansiysk, the provincial seat, to get better acquainted with community and business leaders there.

Amoco had built an operating base at the field site between the cities of Nefteyugansk and Khanty-Mansiysk on the Ob River flood plain. The Ob was one of a series of rivers in Russia that flowed north to the

Arctic Sea. It was midwinter in Western Siberia, a good time of year to visit actually. The rivers had frozen so solidly that they could safely be employed as roads. Just as importantly, the mosquitoes (stories abound that they are so plump that they have difficulty getting airborne following a blood meal!), ubiquitous in the abbreviated summer season, had vanished to wherever the species took refuge during the winter. In the spring, the Ob River was plagued by monumental floods caused by the upper (southern) reaches of the river thawing while the delta, which lay well north of the Arctic Circle, remained frozen and acted effectively as a dam. The Priobskoye field was located about 650 kilometers (400 miles) south of the Arctic Circle.

Still, the weather in Western Siberia was frigid. Lows of −20° to −30°F at night were not uncommon, and temperatures during the day frequently remained below zero. The average high and low temperatures in January in Khanty-Mansiysk, according to WorldWeatherOnline.com, are −16°C (3°F) and −23°C (−9°F), respectively.[1] Of course, plenty of days get colder than that! We were outfitted with specially designed Arctic clothing. Amoco had recently lost a local employee whose car had stalled as he was returning home at night. He froze to death within a couple hours before a rescue team could reach him.

We loaded cases of hors d'oeuvres and other provisions onto a chartered Russian helicopter and set out on the flight to Khanty-Mansiysk, 60 kilometers (40 miles) down the Ob River. Russian helicopters are renowned for their towing capacity and have therefore been utilized from time to time in the international oil and gas industry. When we reached the airfield in Khanty-Mansiysk, the pilot landed the craft but chose not to shut down the rotors. When we questioned why he did not do so, he said that his firm would be charged an extortionary payment if he "landed," which was defined as halting the blade rotation or shutting down the engines.

I glanced down and observed that the tarmac was a sheet of ice below.

One by one, we carefully disembarked and handed down the heavy crates of provisions. As we edged away from the aircraft, we were literally propelled along the ice by the "wind" from the helicopter's rotors in the general direction of the terminal building. I have never been so cold in my life. I envisioned becoming a permanent ice statue on display for the Russians to admire all winter long—an "American negotiator sculpted in ice." What a great art exhibit! The windchill effect was amazing. We all made it inside without serious mishap, and the reception was useful in explaining our proposed project as we continued the lengthy process of building trust within the community.

Meanwhile, the engineers based at the field site in Nefteyugansk were growing restless. They felt that the negotiations were taking far too long and that the best way to convince the Russians of the value of a joint venture was just to begin working together. The agreements could follow later. Perhaps an interim agreement could be signed. They quietly put together a proposal to management to do just that. Support for the proposal ran high among the operational folks, but the commercial and legal teams strenuously objected, pointing out that it would greatly erode the company's negotiating position. For the time being, the commercial forces prevailed, but we knew we had only a brief window in which to complete the deal.

Others were also growing impatient. A Russian, hired locally by Amoco's Moscow office and who had prior (and perhaps current) KGB ties, complained to the head of Amoco Eurasia that I was taking too tough a stand in my negotiations. He used a typical Russian tactic of trying to discredit me. Luckily, the head of the division voiced his support for me, both privately and publicly, stating that it was my job to be tough. That said, I felt pressure rising up like a neap tide on all sides. In reality, the pace of discussions was dependent upon a number of factors, some of which were outside Amoco's control.

Negotiations were largely complete when I left for the treasury position with the company in Chicago. Years later, after BP had acquired Amoco, the British company elected to drop the Priobskoye project. The Achilles' heel of the joint venture had always been that the giant oil field, unlike Sakhalin, was landlocked. A successful project was entirely dependent upon continuous access to an export pipeline in order to transport the oil to distant markets. While a pipeline was not far away, access rights could always be revoked at some point in the future for political or other reasons.

> ### ◎ NEGOTIATIONS PRINCIPLES
>
> One should never commence work on a project before the main agreements are in place, no matter how tempting the perceived short-term benefit. To do so is to sacrifice one's negotiating leverage. The foreign company has as its best weapon the threat of withholding funds; after all, money was what the Russians needed most (expertise was second). If they had had the funds, they would have proceeded without Western assistance. To begin spending on the project in order to showcase our technical expertise would have sent the clear message that, despite any assurances to the contrary, we were not prepared to walk away from the opportunity.
>
> Many arguments were advanced in support of commencing a pilot project, but none outweighed the fundamental erosion of negotiating position and leverage that would dramatically undercut the long-term economics of the project. It was easy for people to forget that the company undertook such projects not to demonstrate technical expertise but rather to make an attractive return for its shareholders.
>
> One of the other lessons we learned in Russia was that even a small local stakeholder in a project carried an inordinate amount of influence since it could easily mobilize the Russian authorities to support its position. The authorities, in turn, could hold up the project in any number of ways until the issue was resolved favorably for the Russian participant.

Back in Moscow, Yuganskneftegaz was undergoing substantial organizational change. Per a Russian privatization decree issued in April 1993, YNG and a Volga refining company, Kuibyshevnefteorgsintez, were brought under the umbrella of a newly created holding company, Joint Stock Company YUKOS.[2] The production association Samaraneftegaz, several refineries in the Samara okrug, and some other assets were also added to the mix.[3] The new entity, YUKOS, was free from direct government control. However, it was still only at the early stages of becoming a Western-style corporation. For example, the share registry was kept in a book, I am told, with entries recorded in pencil no less! Presumably, that facilitated quick changes in ownership with just the stroke of an eraser!

In 1995, not long after being established, YUKOS was acquired by Bank Menatep for less than US $500 million.[4] Whereas the interests of joint venturers Amoco and YNG had been closely aligned at the outset and remained reasonably so once YUKOS was formed, the new owner, Bank Menatep, under the leadership of its chairman, Russian oligarch Mikhail Khodorkovsky, was more interested in extracting money from its holdings than injecting new cash for investments. Khodorkovsky knew little about the oil business when Bank Menatep acquired YUKOS, but he was adept at financial matters and sought to plug the major hemorrhages of cash flow which YUKOS was experiencing. A significant portion of YNG's production was being sold outside of the YUKOS system, and YNG was making massive social payments to the community.[5] Khodorkovsky attended a few of the JV meetings and impressed me with his intelligence, his arrogance, and in some ways, his naiveté.

At the beginning of the next decade, Khodorkovsky locked horns with President Vladimir Putin with disastrous consequences. Reportedly, the YUKOS chairman was considering selling a significant stake in his firm to a Western company, either ExxonMobil[6] or Chevron, and also had ambitious political aspirations at the national level, which Mr. Putin may have felt represented a direct challenge to his increasingly autocratic rule. YUKOS was accused of owing back taxes, and Khodorkovsky was charged with fraud, arrested in early 2003, and imprisoned for a lengthy sentence. He spent nearly ten years cooling his heels (and other parts of his anatomy) in prison in Siberia. Khodorkovsky eventually sought his release on compassionate grounds so that he could see his ailing mother. President Putin ultimately approved the request, freeing him in December 2013, just prior to the 2014 Winter Olympics in Sochi. Prior to his incarceration and a state-forced breakup of YUKOS, Khodorkovsky was the sixteenth richest man in the world in 2004, according to *Forbes*, and the wealthiest in Russia (US $15 billion).[7] He and his management team had built the company into a formidable economic powerhouse.

TEN

Intersecting Destinies
A Near Miss in London

A DECADE AFTER I HAD WOUND UP MY WORK ON THE Priobskoye project, I was involved in one other incident involving the Russians. While living and working back in Houston, I returned home from the office one day to find the message light flashing on my answering machine. Most of the recordings were typically wrong numbers or commercial solicitations. Accordingly, I was tempted to hit "delete" without listening, but of course I did not. I had no inkling when I pressed "play" that I was about to get the scare of my life.

"Mr. Young, this is Dr. Richard Brown of the Centers for Disease Control in Atlanta. We have been notified by the State Department that you recently stayed at the Millennium Hotel on Grosvenor Square in London. We have reason to believe that you may have been exposed to the radioactive substance polonium. Please call me immediately."

The phone nearly slipped from my hands as I recoiled in fear. Notwithstanding the sun, which shone brightly through a nearby window, a deep darkness descended immediately around me. I knew little about polonium but was broadly familiar with the media reports of

Alexander Litvinenko, a former KGB operative and outspoken Kremlin critic, who died gruesomely in a London hospital of presumed radiation poisoning. He had apparently ingested polonium-210 on November 1, 2006, during a meeting at the Millennium Hotel, lost all of his hair, and rapidly wasted away. All the while, his team of doctors could do essentially nothing for him, other than to make him comfortable. The incident sparked international outrage and a major diplomatic row between the UK and Russia. BBC News reported that the UK coroner acknowledged that "Russia could well be involved" in Litvinenko's murder.[1]

While working for the KGB and then its successor organization, the FSB, Litvinenko reportedly had been ordered to assassinate an influential Russian businessman, tycoon Boris Berezovsky, with whom he was well acquainted.[2] Litvinenko allegedly disobeyed the order and then publicly exposed both that operation and others, embarrassing then–FSB chief Vladimir Putin.[3] Putin had him tried for treason and imprisoned for nine months.[4] After his release, Litvinenko and his family fled to the UK where they were granted political asylum.[5] From the relatively secure confines of London, the former KGB operative felt safe to resume his vocal criticism of the Kremlin.

A quick glance at my calendar confirmed that I had indeed stayed at the Millennium Hotel on the night of November 1st, where and when Litvinenko was allegedly poisoned. I was returning home from an employment interview with BG Group. I was ultimately offered the post of head of commercial, which I accepted. At my request, my new boss's secretary booked me into a hotel in Central London for the night prior to my departure for Houston. Her choice of the Millennium may well have been a function of where BG could obtain the best corporate rate. London offered a wider assortment of good restaurants and a comparatively easy trek to Gatwick Airport in the morning, which is why I wanted to stay there rather than in Reading where BG was based.

Before calling Dr. Brown, I researched polonium-210 (^{210}Po) and determined that it was a radioactive isotope that had very limited ability to penetrate (for example, it could not pass through a single sheet of paper or human epidermis), but it was also a high emitter of destructive alpha particle radiation. Accordingly, if it were inhaled or ingested in the minutest quantities, it could do irreparable damage to the human body, usually resulting in death.[6] Some experimental work has been conducted on the effects of polonium on laboratory animals, but to date there is no effective treatment regimen or antidote.[7] It was unlikely that organizations other than state governments would have had access to the substance.

I could feel my heart pounding as I dialed Dr. Brown. After I identified myself, the doctor asked me if I had entered the hotel bar on the evening of November 1st.

"No," I replied warily, fearing what questions lay ahead. "I went to bed early that evening and checked out of the hotel at about 6:00 a.m. the following day to catch a plane back home to Houston."

"OK, good. Your chances of significant contamination would be less then. I would still suggest that you get tested for polonium since the substance was found at various locations throughout the hotel, not just in the bar area itself." Allegedly, it was in the hotel bar that Litvinenko fatefully consumed a cup of tea laced with polonium while chatting with two former KGB colleagues, Andrei Lugovoi and Dmitri Kovtun. When it investigated the crime, Scotland Yard found that samples of the polonium isotope were "off the charts" in the tea cup, and many hotel employees were significantly contaminated with the substance as well.[8] Hence, the authorities concluded that Litvinenko was indeed poisoned in the bar area.[9] That said, traces of polonium were discovered on the Millennium's eating utensils, for example, and I had ordered room service. All in all, not terribly comforting.

"My assistant will let you know where you can get the testing done," Brown continued. I took down the information, thanked the CDC for notifying me, and then ended the call.

I had toyed briefly with the notion of having a drink at the bar, but as fate would have it, I opted instead to turn in for the night, as I was pretty exhausted after my interviews. Still, I *had* stayed in the hotel that night; I *had* breathed the same air as the deceased Russian KGB agent; and I *had* no doubt ridden in the same elevator as Litvinenko had used.

After mulling it over, I decided that there was little point in getting tested since no treatment for polonium poisoning was available. I probably had traces of the substance on my belongings, but I figured I would wait to see if my hair fell out! Fortunately, it did not.

This episode was clearly one that caught my full attention! I marveled at how innocently, unintentionally, and randomly my path had crossed with Litvinenko's. I was so relieved that I had opted to bypass the bar that night.

Litvinenko's widow, Marina, sought and received a favorable ruling from the UK High Court to permit a public inquiry to go forward in order to determine who killed her husband and what role, if any, the Russian authorities may have played. The UK government had originally planned an inquest instead. However, the coroner, Sir Robert Owen, argued that access to key intelligence material, needed to shed light on the situation, would not be allowed in an inquest.[10] One of the suspects, Dmitri Kovtun, has agreed to testify at the ongoing inquiry, continuing to proclaim his and Andrei Lugovoi's innocence.[11] No doubt, more twists and turns lie ahead on the matter.

ELEVEN

It Takes Two to Tangle
Vietnam during the US-Led Embargo

THE YEAR 1990 WAS A PARTICULARLY BUSY ONE FOR ME professionally. Amoco was invited by Vietnam's state energy company, PetroVietnam, to travel to Hanoi to describe the company's technical and financial capabilities. Several colleagues and I were asked to go.

An embargo, which the US imposed upon the Southeast Asian nation in 1975 following the conclusion of the Vietnam War (known locally, not surprisingly, as the "American War"), was still in place. US firms were prohibited from doing business with America's former adversary, and Washington encouraged its allies to follow suit in economically and diplomatically isolating the Vietnamese government.

Although *discussions* between the parties were permitted, US individuals and companies were strictly forbidden from conducting any business transactions with, or conferring any benefit upon, a Vietnamese entity or person. The penalties for violating the embargo were quite severe, both for the individuals committing the offense and for the company for whom they worked. Consequently, as a matter of internal corporate policy at Amoco, an attorney was required to accompany us on any trips to Vietnam to ensure that we did not inadvertently violate the terms of the embargo.

Accordingly, the purpose of my inaugural visit to Vietnam in March 1990 was to conduct business discussions and to develop some rapport with the leadership of PetroVietnam. Offshore blocks were rapidly being awarded to our European and Asian competitors, so we were eager to position ourselves so that we could move quickly once the embargo was lifted. Essentially, our mission was to build a professional relationship and foundation of trust with PetroVietnam. In addition to a lawyer, I traveled with an interpreter, Tran, and a business colleague, Jack, a geoscientist who had been to Vietnam previously. Based in Singapore, Jack was responsible for spotting, screening, and assessing potential new venture opportunities. It seemed quite the enviable assignment to me.

Since the United States had no direct diplomatic relations with Vietnam, we had to travel to a third nation in order to obtain a visa to visit. The process entailed obtaining a letter of invitation from a sponsoring entity in Vietnam that would then be presented together with the foreign passport and the requisite fee for a single-entry visa, usually a one-day process. Since there were nonstop flights from Bangkok both to Hanoi and to Ho Chi Minh City (formerly Saigon) on Thai Airways, my business colleagues and I would typically travel to the Thai capital to obtain our visas. As a fitting bit of irony, the embassies of the former Vietnam War combatants were situated directly across from one another on Wireless Road in Bangkok. Vietnam seemed to delight in stamping a visitor's passport with its chartreuse-colored visa, which was certain to attract attention with the border authorities back home.

The flight on Thai Airways took about 1¾ hours as we traversed the Indochina peninsula northeastward across eastern Thailand and Laos. As I was a US Naval supply officer during the Vietnam conflict, though I never served there, I wondered how I would react to arriving in Hanoi for the first time.

As I sat daydreaming, I reflected back upon an episode from my US Navy days. In 1973, my ship, the USS *Dale (DLG-19)*, a guided-missile cruiser, was set to deploy to Vietnam for six months when hostilities ceased. My boss on the ship told me *never* to run out of underwear to sell in the ship's store on a deployment. He said that an underwear shortage would be the worst thing imaginable that a supply officer could do since it would serve to undermine greatly shipboard morale. Taking his admonition to heart, I loaded onto the ship a six-month supply of Hanes undergarments. It took about 1½ hours for my team to offload the eighteen-wheeler, carry the boxes down the pier and up the gangway, and store them below decks. To my astonishment, the deployment was canceled shortly thereafter, and I spent the balance of my two years on the *Dale* signaling every passing ship to inquire whether they might be interested in swapping my precious (and seemingly limitless) cache of undergarments for more marketable merchandise. That occasion represented a very unusual and protracted negotiation!

On a more solemn note, I had college friends who were killed in action in Vietnam, and I confess that I harbored a certain degree of enmity and distrust toward the Vietnamese. Could I dispassionately view them as potential future business partners, or would I continue to think of them as the wartime "enemy"? The Vietnam War had long ago ended, but the names of places like Dien Bien Phu, Pleiku, Da Nang, My Lai, and Haiphong Harbor came flooding back to me like they had just been on last night's evening news.

As our plane descended for its final approach into Hanoi's Noi Bai International Airport, Jack pointed out several small, perfectly shaped lakes. "Those are not *natural* lakes, my friend, but rather our bomb craters." I looked around me nervously to see if anyone else was listening, but there did not appear to be any other Americans on the flight. I felt that we were already quite conspicuous. Americans had only recently been allowed to travel to Vietnam, and it appeared that my countrymen were not exactly flocking to Hanoi for a holiday! In actuality, the terms of the embargo made such a trip virtually impossible anyway.

"Also, if you look closely, you might see some Russian MiGs on the military side of the air field. The pilots practice takeoffs and landings there." Sure enough, as I peered out the window, I spotted the military aircraft to which Jack referred.

"A representative from PetroVietnam, who speaks English, should be meeting us on the far side of passport control," Jack continued.

As we cleared the immigration formalities, I saw a man in the crowded and noisy lobby hoisting a sign with the familiar Amoco red, white, and blue torch and oval logo. Pressing through the masses, Jack extended his hand and greeted him, "Mr. Lan! Hello! How are you?" Dressed casually in jeans, Lan was rather short and wiry with an uneven and somewhat sparsely populated dark black mustache.

We piled into PetroVietnam's decrepit minivan. At the time, Vietnam was one of the poorest countries on earth, and based upon the ride into Hanoi from the airport, I had little trouble believing that statistic. Scores of barefoot farmers were working the rice fields with water buffalo, and they appeared to live in small, corrugated metal houses. At harvest time, I was told, they would place their rice in the road to be run over by passing vehicles. Absent suitable machinery, the process was intended to pry the rice free from its husk. A pragmatic approach, I suppose, but not particularly sanitary. Despite the prevalence of poverty, Vietnam could boast a literacy rate of nearly 90 percent among its sixty-seven million or so inhabitants at the time.[1,2]

Lan, who from his perch next to his driver in the front seat had been conversing quietly with Jack, turned his gaze to me. "First time in Vietnam, or were you here during the American War?"

"No," I quickly replied, almost defensively. "This is my first visit. Beautiful country." After I said that, I realized that my observation would be seen as rather shallow and insincere since I had just landed and clearly had had no chance to view the country thus far.

We ascended a ramp to access a major bridge across the Red River. Twisted remnants of two similar structures lay in ruin on either side of us, and all three bridges were badly rusted. Hanoi was situated just beyond an embankment or berm on the south side of the Red River.

Observing me staring at the wreckage, Lan chuckled. "Do you know why this bridge is intact while the others have been bombed into junk?" Without awaiting a reply, he added, "We lined up your servicemen in a human chain across this bridge so that your pilots would not bomb it."

I felt my heart racing as I tried to control my inner rage. I thought, *What a great way to start a relationship with visiting Americans!*

Hanoi was an amazing place, bustling with bicycles but comparatively few motor vehicles, a bit like Beijing in its early days. As dusk descended, it was amazing how few electric lights were visible. Kerosene lamps illuminated many homes. It appeared that the US-led embargo had had the intended result of stifling Vietnam's economic growth and development. With the lack of light and periodic shortages of electricity, the city felt as though it had been left to wallow in the early twentieth century, largely doing without modern conveniences.

I observed that businesses of a given type (e.g., barbers) were grouped together in a single section of the city. I guess state planning had its benefits—one did not need to travel to widely separated districts to compare and contrast different barbers! It was also interesting to note, despite the lack of many other electronic devices, how many residences had Japanese-branded TVs. Given that Tokyo was "officially" adhering to the US-led embargo of communist Vietnam, I found that fact surprising. It appeared that the Japanese government's official support of the embargo was, in fact, a bit disingenuous, or at least they were not rigorously enforcing it. I supposed that the TVs could have been sold legitimately to merchants in a non-embargoed nation like Indonesia, which then could have sold them on to Vietnam.

We passed by the historic core of Hanoi with its French colonial buildings painted in pastel shades of yellow and pink. Although they could use a fresh coat of paint, they generally looked to be in a good state of repair. Shortly thereafter, we came upon a large tranquil lake almost completely shrouded in mist. Along the side of the road, Lan pointed out a monument to the capture of an American pilot. It sat about 25 feet away, largely obscured in the gathering fog, barely visible with the help of a few low-wattage street lights. "This is West Lake. We rescued your pilot from the water," he explained. Jack added that the monument was to the downing and capture of Sen. John McCain who parachuted into the water nearby. The Vietnamese claimed to have "saved him from drowning," and perhaps they did. However, knowing the brutal torture to which he was subsequently subjected, I felt my anger boiling up anew.

We pulled up to the Thang Loi Hotel, a white sprawling complex that, upon closer examination, comprised a main building, where the lobby was situated, and a series of bungalows on stilts that extended out into the shallow water. The bungalows were connected by a labyrinth of elevated wooden walkways. Lotus plants, many in bloom near the shoreline, were illuminated by spotlights.

This would be our base for the next few days. The Thang Loi, Lan informed us, was a gift from the Cubans. I was told that PetroVietnam placed their guests either in this facility or in government guesthouses that had been constructed by the French. The guesthouses, where I lodged on subsequent visits, were beautiful French colonial structures, each equipped with an underground bunker into which occupants would clamber when air raid sirens sounded during the bombing by the Americans and their allies. Although paint was chipping off of them from years of neglect and they could have benefited greatly from a power washing, they otherwise were in remarkably good shape and seemed to have escaped any direct hits during the bombing campaigns.

Our passports were taken from us upon check-in at the front desk of the hotel. Jack explained that this was standard operating procedure.

We would be "registered" with the local authorities who were the same people that would grant permission for us to leave the country at the conclusion of our visit. "You will get your passport back in a couple of days," Jack assured me. I was a bit uneasy about being in Vietnam sans passport, but my angst turned out to be misplaced, and the process, in fact, worked rather smoothly.

Rooms at the Thang Loi were reasonably comfortable, though sparsely outfitted with just a few wall hangings. They had the typical Soviet-style layout of two single beds separated by a small bedside table. A straight-backed chair was positioned by the lone window. The room was incredibly basic but did have a nice view out over the strangely serene West Lake. The hotel staff was exceedingly friendly.

Not knowing what the food would be like, I had brought a stash of cookies, cheese, and breakfast bars. It seemed a prudent measure until one day I found some of those items relocated from my suitcase, which was sitting open on one of the beds, to an area under the bedside table. The food appeared to have been partially consumed. A can of pressurized cheese spread had been punctured, squirting cheese around in the vicinity. It seemed that an animal had been torn between the twin objectives of building a nest and having lunch.

Then I spotted it—an oversized rat in the far corner of the room. I grabbed the phone and dialed the front desk for help (or perhaps, HELP!). They promised to send housekeeping right away. I surmised that mine was not the first such call. As I was already late for my next meeting, I grabbed my attaché case and headed out the door (rapidly, I might add). As a consequence, I have no idea whether anyone actually came following my departure.

When I returned later that day, I found no sign of my "roommate." However, the next morning I either saw the same critter or his twin scurrying about the hallways of the Thang Loi, appearing smug and perhaps a bit plumper. He must have received quite a shock when he gnawed into the pressurized can of cheese! I have to say, I slept fitfully after that encounter, fearing that the rat would bite me, prompting the need for a series of rabies shots. Not a fun experience, I am told, since they are administered to the stomach. The thought of visiting a local doctor was equally unappealing.

Our informal discussions with PetroVietnam focused, as you would expect, on potential future business collaboration. I had prepared a fact sheet on Amoco that I reviewed with the meeting attendees. I knew that they would appreciate a document in their language. Accordingly, I had had it translated into Vietnamese prior to leaving Houston.

"I see that this was prepared by someone from the South," noted Mr. Han, another PetroVietnam official at the meeting. It had not occurred to me that a) there was still plenty of residual animosity toward southern Vietnam or b) the language varied significantly and therefore was easily detectable.

Apparently, the dialect is sufficiently different so as to make the source readily apparent. The word for "oil" in the South actually meant "kerosene" in the North. So the name "Amoco Vietnam Oil Company" became "Amoco Vietnam Kerosene Company," not at all what we intended and not descriptive of the business we wished to pursue in the future. I apologized for my oversight. As I reflected on the blunder, it occurred to me that virtually every Vietnamese translator in the US would have been a descendant of the South Vietnamese who fled their country in 1975.

As the meeting continued, PetroVietnam wanted to know when the US embargo would be lifted. We explained that we had no control over the process and would have considerable difficulty in forecasting that major milestone. Lan leaned over and handed me an envelope. "Please deliver this message to your State Department." I almost instinctively launched into a mini-lecture on the difference between the private and public sectors in the US and that our company was in no way connected to the government. This was a speech which was well rehearsed from my trips to Baku. It was obvious that my explanation was falling on deaf ears, however, so I reluctantly agreed that I would send the letter to Washington upon my return. The Vietnamese were clearly getting few American visitors and were starved for attention.

As it was my first visit to the Vietnamese capital, I was "invited" to take a tour of the city, which consisted of visits to Ho Chi Minh's mausoleum, one of the war museums (which displayed parts of downed US aircraft), and a drive through a part of Hanoi where many of the disabled from the war resided. Both sides in the protracted conflict suffered massive casualties. A long line of schoolchildren, tourists, and well-wishers were queued up to view "Uncle Ho," as he was affectionately referred to locally. PetroVietnam took us to the front of the line, explaining to the somewhat disgruntled masses that this was a state visit (well, hardly!).

Ho Chi Minh looked surprisingly vibrant and alert, with just the hint of a smile on his lips, as he sat propped up in his bed. Lan, who accompanied us, explained that Ho Chi Minh's body was transported to Moscow periodically for "touch up" work. He seemed well informed and current on all things Vietnamese. I later was told by a reliable source that Lan had been a Saigon-based spy for the Viet Cong during the war. If that report was accurate, I wondered how many American deaths resulted directly or indirectly from his espionage activity. US forces frequently complained to their superiors that the North Vietnamese army seemed to know with precision accuracy the American battle plans in advance and were accordingly well prepared to fend off any attack. I suspected that there were many more like Lan during the war who operated undetected since it was so difficult to determine which Vietnamese were friends and which were foes.

It was summertime in Hanoi, and daytime temperatures were routinely over 32°C (90°F). The local joke was that the only air-conditioned space in all of northern Vietnam was Uncle Ho's mausoleum! When our meetings and tour had wrapped up, Jack invited me to join him for a beer in the bar at the Thang Loi.

"What will it be? The options are 333, locally brewed and really not too bad, or Old Milwaukee."

"What? Are you serious? Old Milwaukee?" I reacted incredulously.

"Yes, indeed I am. The Viet Cong apparently captured a container full of our beer, and the Vietnamese have been selling it back to us ever since. They may be Communists, but they seem to understand the basic principles of capitalism just fine."

Dinner was at Restaurant 79. The few eating establishments on Amoco's unofficial list were referred to by number (except for the Piano Bar, which was self-explanatory), apparently based upon their location on a given block. As many of the blocks looked virtually identical, I wondered if a solitary number would be sufficient to describe the exact location of a given restaurant. Perhaps there were relatively few deemed suitable for foreign visitors.

The voyage to the restaurant that evening was particularly memorable. It had been an unbearably hot day, accompanied by high humidity. I was accustomed to this combination from my years of living in Houston, but somehow this day seemed far more oppressive than a summer day in the Bayou City, probably due to the shortage of air-conditioned facilities. I began to see flashes of lightning on what I reckoned to be the western horizon. Soon we were ensnarled in a monumental traffic jam—a veritable gridlock—and a ferocious thunderstorm descended upon us. A flash flood rapidly ensued, enveloping the street and surrounding curbs in muddy water. Litter of all sorts floated past us. I feared that the trusty PetroVietnam van would stall out, but mercifully it did not.

Ultimately pulling up in front of the restaurant, we were faced with a dilemma. If there were any sewers, they must have been clogged, which meant that the floodwaters would likely recede very slowly. Accordingly, our choices were either a) to ruin our business shoes wading through the high waters, or b) to remove our shoes, roll up our trousers, and risk contracting an infection as we waded barefoot through the turgid waters. We chose the latter option, which, with hindsight, was probably the less advisable one. I was glad that my tetanus and typhoid shots were up to date!

The fare at the restaurant was surprisingly good, including fresh seafood, snake (yes, it actually *does* taste a bit like chicken), and for dessert, bananas flambé. While the French had long since departed Vietnam following their defeat at Dien Bien Phu, they seemed to have left a lasting impression on the eateries, each of which seemed to offer this French delicacy. The waiter brought the snake cage by in advance of the meal for us to handpick our victim. I pointed at one, having absolutely no clue what would differentiate one snake from another in terms of taste! I glanced at the snake cage again as we left the restaurant and noticed it was abuzz with activity. The reptiles were in a state of agitation, apparently annoyed that one of their own had been sacrificed for the benefit of a few hungry Americans!

Jack explained that there was little available refrigeration in Hanoi and hence items like fresh seafood were trucked in daily from Haiphong Harbor. One day at Restaurant 202, following a lunch, I observed a crab being cracked open on the sidewalk (not by a person merely seated there, mind you, but instead a worker wielding a hammer smashing the shellfish directly on the stained and dirty concrete walkway). I confess that, although I frequently contracted common travelers' maladies on international business trips, I never was sickened by the food during my periodic visits to Vietnam. The seafood, in particular, was delicious.

It would not be an exaggeration to say that evenings in Hanoi were quiet. There was no TV—at least none that we could understand—and, unless we had made arrangements in advance with PetroVietnam, we did not have access to a car and driver.

One evening our interpreter, Tran, suggested that we go to "International Night" at a club within walking distance. He heard that they had live music, drinks, and dancing. It was about a mile's walk from the Thang Loi. International Night was nothing short of amazing. The facility was quite crowded with Vietnamese men and women dressed up as they believed would be the style in the outside world, though their only contact with life beyond the "bamboo curtain" was undoubtedly through interaction with visitors to Hanoi. In most cases, the fashions the Vietnamese wore fell far short of the mark. The women had put on Western-style makeup, but it seemed misapplied and overdone. The event almost seemed like a high school prom or a college formal where nobody was dressed quite right. Still, it was nice to get "out on the town," and it was fun to watch the locals dance to songs like the "Lambada." The last thing I expected to see was Brazilian influence in a club in Hanoi, half a world removed from Rio.

The evening of "people-watching" passed by all too quickly, and we realized that we needed to get back to the hotel. As we went to leave, a beautiful young lady suddenly ran up to our lawyer, Michael, hugged him, and pleaded for him to take her to the US. As I examined her more closely, I realized that she was an Amerasian. Her American father or mother had unintentionally—or perhaps *with* intent—left her behind when returning home to the US after the North Vietnamese overran the South and the US withdrew its forces. It was an awkward scene to say the least. Michael pried himself loose and we quickly strode away with the sounds of her wailing growing fainter as we put some distance between us and trouble.

As we made our way back to the hotel, we discussed the Amerasian lady's plight and how we had narrowly averted an international incident. What if the police had been there and felt that we were plotting to spirit her out of the country? How could we prove that we were innocent of any wrongdoing?

Partway through our conversation, Tran chimed in with his own story. In the early 1970s, he was a pilot in the South Vietnamese Air Force, flying daily bombing missions over the North. As the North Vietnamese forces pressed south, and it became apparent that the South Vietnamese regime's days were numbered, Tran arose one morning, suited up as per usual, and climbed into the cockpit of his fighter. But instead of heading north, he steered his aircraft southwest to Thailand and somehow made his way on to the US where he sought and was granted political asylum. We told him that, had we known his personal history, we would never have invited him to participate in our trip to Vietnam. We felt he had taken an extraordinary risk. Tran would not accompany us again. There were undoubtedly many stories like his, men and women taking huge risks to escape to freedom and an opportunity to begin a new life. Sadly, for all who escaped successfully, there were family members who stayed behind to face what wrath the government might unleash in their direction.

Indeed, after the war ended, many in the south of the newly unified country were oppressed. Some South Vietnamese, feeling that the potential rewards outweighed the risks, set out in overloaded and poorly constructed boats for distant shores like Hong Kong. Some never made it to their intended destinations; their boats were ill-equipped for lengthy sea voyages. Others were refused entry upon their arrival, herded instead into hurriedly prepared detention camps with deplorable living conditions.

⊘ NEGOTIATIONS PRINCIPLES

As mentioned earlier, successful negotiations depend upon establishing mutual trust. In the case of Vietnam, I worked to lay the groundwork for eventual negotiations. However, because of the US embargo, our activities were limited strictly to discussions. This created a particularly difficult challenge in Vietnam. We wanted PetroVietnam to appreciate that we were a results-oriented and dynamic company, yet all we could do was talk longingly about the future. If we visited Hanoi too often, we would quickly wear out our welcome. On the other hand, if we did not come frequently enough, we feared that we would become "out of sight, out of mind."

The competitive clock was also ticking on Amoco as PetroVietnam, led by Lan and his colleague Han, issued one announcement after another that it had awarded various blocks to firms from other countries. I was struck by PetroVietnam's cleverness in spreading the offshore acreage among various countries and companies, many of them European. In that way, no one entity could become too influential. Clearly, they understood the benefits of competition. They were reportedly receiving consulting advice and assistance from Petronas, the Malaysian state energy company.

One of the most difficult aspects of the embargo had been the restriction on in-country expenditures. A visitor to Vietnam was limited to spending no more than US $100 per day. On one occasion, I needed to place a call from Vietnam to a colleague in Singapore in order to separately relay a message back to our US office (calls could not be placed directly to the US because then an American company would be doing business directly with a Vietnamese entity or the state). The cost of the call was US $90, which left me with a mere US $10 for meals for the balance of the day.

Following that trip, I traveled to Washington to appear before the Treasury Department's Office of Foreign Assets Control, the entity that administers and enforces the US embargo programs, to testify in support of an increase in the per diem rate. I explained the hardship that the spending cap was causing Americans traveling to Vietnam. Shortly thereafter, the rate was raised to US $200 per day.

Young and Amoco colleagues on the streets of Hanoi

I made a number of similar relationship-building trips to Vietnam. One of them was to Ho Chi Minh City in early 1991 when I stayed at the Caravelle Hotel. During the Vietnam conflict, the Caravelle was considered the lodging of choice, I was told, of celebrated foreign correspondents like Dan Rather and Peter Arnett. "On every big story I have covered these past fifty years," Arnett recalled, "there has always been a favorite press corps hotel, from the Constellation in Vientiane in the late 1950s, to the Caravelle or the Continental in Saigon . . . to the Al Rashid in Baghdad in the first Gulf War to the Palestine in the second. These places became our homes through shelling and sieges, and where we bonded or broke up."[3]

Early one evening after work, I took a stroll past the former US Embassy compound that still looked much as it did that fateful day of April 30, 1975, when the US pulled out its remaining personnel from the embassy as North Vietnamese forces descended on central Saigon. Eerily, the facility now seemed somehow to be in a deep slumber or in suspended animation. Reportedly, PetroVietnam was using the building to house its stash of oil and gas seismic tapes since the air-conditioned facility offered some protection from the stifling heat and humidity outside. I also toured the "Reunification Palace," which was the headquarters of the South Vietnamese government prior to 1975. Now a museum, it was easy to picture its past as the presidential palace, South Vietnam's governmental leaders scurrying around tending to matters of state and entertaining legions of visiting foreign dignitaries.

During this visit to Ho Chi Minh City, several business colleagues and I paid a courtesy call on VietSovPetro, a Russian-Vietnamese exploration and development joint venture, in Vung Tau. Located about 100 kilometers (60 miles) southeast of Ho Chi Minh City on the South China Sea coast, Vung Tau must have been highly sought after by Russian oil executives weary of the unrelenting snow and bone-chilling cold of a Moscow winter. As it turned out, one of the Russian officials with whom I had had previous dealings in Moscow was also in Vung Tau at the time. Recognizing each other, we exchanged brief pleasantries.

"Yes," he said as a wry smile crept over his ruddy face, "someone had to . . . how do you say? . . . check on our Vietnamese investments! It was convenient for me." He seemed almost embarrassed to have been caught red-handed (no pun intended) in Vung Tau in February.

Vung Tau was a pleasant seaside resort. I could only imagine that, absent the embargo, it and other Vietnamese coastal villages might well have attracted Western vacationers in droves. My mission was to gain an increased understanding of the challenges of working in Vietnam, to update our knowledge of the joint venture's status, and to determine if they might welcome another partner at some point in the future. I was received cordially, though they took pains to remind me that American companies were barred from working in Vietnam at that point. I responded that I was all too aware of that reality.

We had rented a car for the trip to Vung Tau from Ho Chi Minh City.

About halfway back, the engine stalled, and several attempts to restart it produced only a sickly sputtering sound. I began thinking about what alternative transportation we could find if the car could not be restarted. The thought of summoning AAA, while appealing, was clearly a mere fantasy in Vietnam! None of us spoke Vietnamese, and only a couple any Russian. Fortunately, the story did not end there in a baking hot village in the middle of nowhere. It would be an exaggeration to say that the engine "whirred" back to life, but it did reluctantly turn over, and we made our way uneventfully back into Ho Chi Minh City.

Half a world away in Iraq, the first Gulf War was going on during this trip, specifically Operation Desert Storm, and I remember the receptionist at the Hotel Caravelle's front desk remarking rather ironically as I checked out of the hotel, "I see you have had better luck with *this* war!" Nothing like rubbing it in!

Upon my return to the US, I presented my passport to the customs and immigration officials. The agent looked up and began interrogating me.

"What business are you in and where is your home?" he asked.

"The oil business and Houston, Texas," I replied matter-of-factly, wondering whether he felt I had somehow inadvertently violated the Vietnamese embargo.

"That does not add up," the agent continued.

"Why?" I inquired with growing unease.

"If you were really from Texas, you would have replied that you were in the 'awl bizness.'" He grinned, and I realized that I had managed to find the one US border agent with a sense of humor.

"You have me there, sir."

While attending a conference in Singapore in 1992, I was approached by a business developer from BHP, the Australian company, inquiring whether Amoco might be interested in partnering in offshore Vietnam. I politely declined, explaining that the US embargo remained in place.

As it turned out, that timing was fortuitous for the company. During the embargo years and the decade thereafter, only a few relatively small discoveries resulted from the exploratory drilling off the Vietnamese coast, and most of those proved to be uneconomic. Were it not for the embargo, we almost certainly would have farmed into one or more of those projects. On this occasion, it seemed that we had been saved from ourselves.

Although Amoco and many other American companies actively lobbied the US government to lift the Vietnam embargo, the companies ran into stiff resistance. The US foreign policy was to continue to isolate Vietnam. Bill Clinton lifted the embargo on February 3, 1994. Soon thereafter, Amoco opened an office in Hanoi.

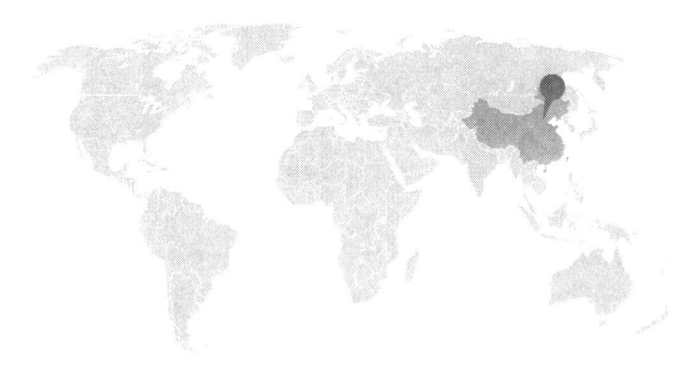

TWELVE

Sea Slugs, Mao Tai, and Intense Negotiations

CHINA IS OFTEN DESCRIBED AS INSCRUTABLE, A CULTURE that confounds and mystifies even those outsiders who have spent a lifetime studying it. My exposure to the People's Republic of China (PRC), prior to my first visit, had been confined to reading books like Pearl S. Buck's *The Good Earth* and consuming Chinese food. I confess that I was among the naive and ill-informed who believed that fortune cookies were genuinely Chinese and not an American marketing gimmick (I was wrong).

I distinctly recall my first visit to Beijing in the mid-'80s as a project economist for Amoco. I stayed, as many foreigners did, at the Sheraton Great Wall Hotel, one of the few facilities deemed suitable for Western business travelers. From my hotel room, I had an unobstructed view of a park four stories below. Early in the mornings, I would see both the young and the venerable striking strange poses as they practiced their *tai chi*.

I marveled at the bicyclists, who greatly outnumbered cars at the time. At rush hour, legions of them would steer their bikes effortlessly through an intersection, managing to glide past other cyclists coming from both the left and the right. It was akin to watching a flock of birds that routinely fly through another flock, somehow averting collisions. The birds seemed to possess onboard radar or some other collision avoidance software. So it was with the Beijing bicyclists.

When bicycles were the primary conveyance in Beijing and other Chinese large metropolitan areas, air pollution was far less an issue than it is today. A study published in the British Medical Journal suggested that residents of the capital are shortening their lives by up to sixteen years by habitually breathing Beijing's choking smog.[1] The topography of Beijing, which is partially encircled by lofty mountains, conspires to exacerbate the pollution problem by trapping the particulate matter emanating from road vehicles and factories.

I am probably one of the few people whose first encounter with a Peking duck was actually in Beijing! It seemed that no part of the bird went to waste in China. The external layer of fat, discarded in many places as unhealthy, was highly valued there. In general, I found the food in Beijing to be delicious, though quite different from what I quickly came to appreciate was actually Chinese-American cuisine back home. I have always been an adventurous eater, and I found the variety of food one encounters on foreign business trips to constitute one of the great "perks" of being an international negotiator.

During the 1980s—in fact, from 1978–1992—China was led by reformist Deng Xiaoping, who had embarked on the modernization of China in the post-Mao period. He took the country down the path toward a market economy, which he termed the "socialist market economy,"[2] and is credited with sowing the seeds that allowed China to become the economic powerhouse that it is today. It was a balancing act of introducing some Western economic principles while retaining the Communist Party's unrelenting grip on authoritarian power. Stating that there was no roadmap for economic reform, Deng implored the Chinese people to "cross the river by feeling out the stones with their feet." To critics of his economic policies, he urged pragmatism—"It does not matter whether a cat is black or white, as long as it catches mice."[3]

Amoco had won the right to explore in the waters of the South China Sea off the southeast Chinese coast. It was my job as a project economist (my role at the time) to persuade my counterparts that the commercial terms we had proposed were equitable and fair and in the interest of our partner, the Chinese National Offshore Oil Company (CNOOC), and the PRC government. We would give CNOOC our computer models, having previously stripped from them sensitive proprietary elements like price forecasts, and we provided some instruction to CNOOC in how to use them. We found that sped up their evaluation process and gave us some comfort that they were properly calculating present values and internal rates of return, two economic measures commonly used in the oil and gas industry (and beyond) to judge the expected profitability of prospective projects. In our analyses, we tended to increase our projected operating costs by a factor to reflect the lack of oil and gas infrastructure and facilities in China at the time and the fact that we would be required to use local contractors for many goods and services. Chinese drilling contractors were still in their infancy back then and operated somewhat inefficiently. The added costs would be passed along to our company.

For our meeting, we gathered in a huge conference room with a long, rectangular table. The Chinese delegation outnumbered us by a ratio of five to one. I recall our delegation leader, Mr. Blanton, partway through the meeting declaring that Mr. Young and the Chinese economist should huddle in a separate room where they could compare models and analyses. As I arose, so did no fewer than five Chinese economists, toting their abacuses. I could see that this was going to be a long day. CNOOC was anxious to learn all they could from their American guests, both technically and commercially, and they were indeed quick learners. They insisted that our agreements include a provision for "shadows," CNOOC employees who would closely follow their Western counterparts and observe all that they did.

I always felt that a thorough understanding of project economics should be a prerequisite to becoming a negotiator. After all, how could one bargain effectively if one did not understand what the various components of a project were worth in present value terms? In Western cultures, companies depended to a considerable extent on the project economics generated by their experts. I was less sure that would be the case in China where optimizing profits might be less of a core objective than carrying out government directives and meeting targets in the state's five-year plans.

I quickly learned that while political mandates played a key role as drivers for the PRC negotiating teams, the Chinese also were very motivated and adroit at extracting every bit of economic rent they could from their Western counterparts. While they might have been lagging in technical expertise and energy industry experience, they more than compensated for that shortfall with their ability to negotiate. In fact, the vast majority of the Chinese with whom I dealt proved to be masterful negotiators. In a typical bidding environment for an offshore block, they would allow a few companies to pass the initial threshold test. They would then declare that these companies had "an opportunity to improve their bids" with no real guidelines provided. Predictably, the companies would scurry off to do just that, acutely aware that they were still in a keenly competitive environment. The ultimate winner would discover that more negotiations lay ahead as the actual PSC terms were considered. With time, the Chinese developed their own distinct model agreement, which was a combination joint venture and PSC. Among the provisions was a requirement to use certain local service providers.

That CNOOC could stir up such fervor and anxiety among the competitors was partly a testament to their effectiveness as negotiators. Contrary to initial lofty expectations, the industry as of 2015 had not found huge reserves of oil and gas off the Chinese coast. Still, the PRC was recognized as an immense and developing market that held incredible potential in the downstream sectors of the energy business—manufacture of petrochemicals, sale and distribution of refined products, and liquefied natural gas

(LNG) importation and regasification. Even in the 1980s, it was clear that China, with its double-digit GDP annual growth rates, would eventually be one of the world's largest energy consumers. Amoco had a relatively small offshore discovery in the Pearl River Mouth Basin, called Lihua. The block we were negotiating on at this time was a look-alike to Lihua—on trend and in close proximity. Oil discovered in this part of the world was frequently waxy, which complicated production techniques and added further to the costs of extraction and processing.

The Chinese were gracious hosts who insisted upon long-standing business traditions, one of which was the banquet. The PRC host would throw a lavish welcoming dinner upon the arrival of the foreign delegation. The guests would be expected to reciprocate just prior to their departure. Typically, there would be a head table at which the dignitaries from both sides would be seated. Circular tables, each fitted with lazy Susans, would be positioned throughout the rest of the banquet venue. Hosts and guests would sit in alternating seats so that the host could serve the visitors from the vast assortment of exotic dishes on the lazy Susan. As I have mentioned, I consider myself an adventurous eater and have always relished the chance to try new and exotic dishes. Opportunities abound in China!

I was advised that it was considered crass to inquire into the specific identity or composition of what we were served. It also might embarrass the host if he or she did not know what the item was. Loss of face is never a good position in which to place your Asian host. Luckily, the gentleman seated next to me spoke a bit of English and passed along the name of each delicacy as he thrust generous servings onto my plate. I must say that I was not a great fan of the hundred-year-old eggs (they taste as though they have been sitting around idly on a shelf for a century or so; they certainly must be an acquired taste!) and the sea cucumbers (also referred to more descriptively as sea slugs). The sea cucumbers reminded me of the layer of blubber one removes from the outside of a ham shank prior to serving it to guests. It had the consistency of jelly and the appearance of fat. In fact, my dislike of sea cucumbers was legendary among the Amoco delegation, and in most cases, they shared my abhorrence. As I was just recovering from a gastrointestinal ailment on this occasion, the thought of the miserable sea creatures particularly nauseated me! Still suffering the ill effects of jet lag as well, I felt a bit dizzy staring at the lazy Susan as it twirled before me. I wondered if the centrifugal force was going to toss some of the containers outward onto my lap. I found myself hoping that, when the wheel finally ground to a halt, the sea cucumbers would not wind up in front of me. The entire process was reminiscent of Russian roulette, Chinese edition—with enough spins of the wheel, everyone eventually loses!

On this particular occasion, the chief negotiator for the project, Mike, was seated at the head table, and the toasts had begun in earnest. Just as the Russians had used vodka to annihilate their inexperienced and unwary visitors, the Chinese weapon of choice was Mao Tai or Bai Jiu, a type of potent liquor that smelled and tasted a bit like kerosene. As I heard a fork being tapped on a glass, I looked up to see Mike rising to his feet. "Ladies and Gentlemen," he began, "as soon as Mr. Young downs his sea slugs, I would like to propose a toast."

Once again, I was in my customary position of "taking one for the team," only to be rewarded thereafter by having to consume yet another glass of Mao Tai. The sea cucumbers slithered disgustingly down my throat as I felt all sets of eyes in the great hall fixated upon me. As I consumed the appetizer, I began to consider what retribution would be appropriate for my friend (former friend?) Mike! My guest offered to refill my plate with the disgustingly slippery little things. I motioned him away, perhaps a bit too vehemently! While his English was marginal, I think he understood "not on your life, buddy, if you know what is good for you" in hand gesture equivalent.

⊘ NEGOTIATIONS PRINCIPLES

The Chinese with whom I dealt were accomplished negotiators indeed. To great effect, they knew how to blend the presence of competition, real or conjured up, with encouragement to "improve one's bid." When companies respond to such a request (versus responding to a legitimate counteroffer), this is referred to as "negotiating with yourself." As discussed earlier, it is far preferable to receive a concrete and specific response from the counterparty, which then establishes where the goalposts are positioned on the playing field. However, in a formal bidding process run by CNOOC, no counteroffer is made at the qualifying stage, so foreign companies must play by the Chinese rules. If they do not show enough movement, the companies are disqualified from the next round. If they move too much, they leave money on the table, undercutting their project economics and leaving them less leeway to compromise at a later stage of the negotiations.

The Chinese also use time to their advantage, knowing that if they establish a deadline and use dilatory tactics to run the foreign company up against that deadline, they frequently can extract additional concessions. On some occasions, we were able to use the same technique to advance our cause. For example, we would set a date for a signing ceremony, to be attended by high-level dignitaries, and then use time pressure to our advantage to prompt some movement out of the Chinese as well.

As I have mentioned, sometimes companies needlessly and inexplicably impose time pressure on themselves. If there is not a similar environment on the other side of the table, the party who has been subjected to the time pressure, even if self-imposed, is far more likely to make concessions in order to get to the finish line more rapidly. This behavior is usually not conducive to obtaining an optimal negotiations outcome.

In a negotiation, it is sometimes possible to identify a component that is disproportionately valued by one party versus the other. For example, if the Chinese objective of training its staff in Western techniques was more highly valued by them than the corresponding cost of the training provided, then the counterparty may be able to extract additional value elsewhere (for example, perhaps a greater share of the profit oil). In this manner, it might be possible to craft a win/win scenario in which both parties would feel they got a better deal.

Opportunities sometimes exist to expand the economic pie itself. Perhaps the Western company can lobby the government for a tax holiday. In exchange, a portion of that economic rent could be shared with the Chinese partner. In such case, the total economic rent available to allocate between the parties has grown to both parties' benefit.

While supporting the negotiations on the China offshore block, I came to appreciate the usefulness—actually, the indispensability—of a present value table that showed what each concession would cost us (or conversely, the extent to which Chinese concessions would benefit us). The table allowed the negotiator to approximate the impact on the economics of changes to a given term. Typically, variances in production rates or retained profit oil have larger impacts than corresponding changes to operating costs. As noted, we were concerned that operating in China might prove exceedingly costly and therefore tended to assume costs would fall in the upper range of estimates. A present value table works more effectively than one for internal rates of return since the latter economic parameter is nonlinear, making extrapolations over lengthy intervals more risky.

In trying to close a negotiating gap to reach a deal, sometimes it is possible to use the time value of money to advantage. Since oil and gas projects frequently have lives that extend from twenty to fifty years, one can provide a counterparty with a disproportionate amount of profit oil out in the distant future since in present value terms those "outer years" will have considerably less impact on the project economics.

> Also, since it is normally the international energy company that has a very large investment at risk, it is clearly better to recover the costs of that investment as soon as possible and to capture the majority of the expected return early on. The longer an investment sits at risk, the greater the chance that an adverse event (e.g., newly imposed taxes, abrogation of agreement terms, or outright governmental expropriation) will occur. Companies frequently insert economic equilibrium clauses into their agreements to increase the likelihood that they will realize the fruits of their negotiations. An adverse modification to a key term would need to be offset by a corresponding change somewhere else in the agreed terms to maintain the same level of economic benefit. Rates of return are rarely guaranteed, and companies appreciate that they are taking technical, commercial, political, and other risks, but they want the chance to earn the rate of return for which they bargained.

Negotiations with CNOOC almost always meant traveling to the Chinese capital, but venturing halfway around the world allowed for some great sightseeing opportunities during brief periods of downtime such as weekends. The Forbidden City, the Summer Palace, the Temple of Heaven, and the Great Wall were all awe-inspiring, even on multiple encounters. I also found the smaller things quite amazing. On a bus ride in the mid-'80s to Badaling, the closest Great Wall site to Beijing open to visitors, I noticed a road crew constructing a cobblestone road by hand, one stone at a time. While it seemed to me to be an astoundingly inefficient and slow process, I realized that where labor costs were low and the workforce so immense, labor-intensive construction methods were more viable than they would be in higher-wage environments such as the US and Western Europe. I suspected that as standards of living and wages rose in the PRC, these old ways would gradually disappear.

One evening during a stay in Beijing, one of Amoco's senior executives from the Chicago corporate headquarters issued an invitation to all the company personnel in town to join him for dinner at one of his favorite restaurants. Such an invitation was considered a "command event" from which there was realistically no escape. Besides, I figured I might expand my repertoire of Beijing eateries. We gathered there at the appointed hour of 7:00 p.m. and were seated at a large circular table. Our host arrived a few minutes later, striding across the room and proclaiming to all in attendance, "Welcome to my favorite sushi restaurant in Beijing! Let's start off with my all-time favorite—lobster sushi."

Immediately, alarm bells sounded in my head. Even the Japanese, who I recognized as the globally acclaimed sushi experts, always served their shellfish cooked. Also, China probably did not have the same stringent food-handling requirements and cleanliness standards as those employed in Japan. While I was contemplating the inevitable, the waiter brought in what appeared to be about a ten-pound Asian lobster, writhing in his hands as if he were repentant for whatever crime he had committed, was contemplating his sentence, and was fully aware of the fate that awaited him. Then the waiter/executioner was off to perform his evil deed. About five minutes later, he reemerged beaming with an oversized platter of lobster sushi (yes, still completely raw) in the center of the platter with the head and tail tip displayed at opposite ends in an artistic effort to reassemble the creature. The lobster's antennae were still twitching from its recent demise. Personally, I love lobster and other shellfish, but eating them raw was definitely not to my liking.

Miraculously, my gastrointestinal system emerged unscathed from the incident. Pressure from both a superior and peers can make it difficult indeed to wiggle out of this sort of situation. Perhaps I could have feigned seafood allergies, but I did not want to be untruthful. In these situations, there was no negotiating—clearly a case of disproportionate relative power levels!

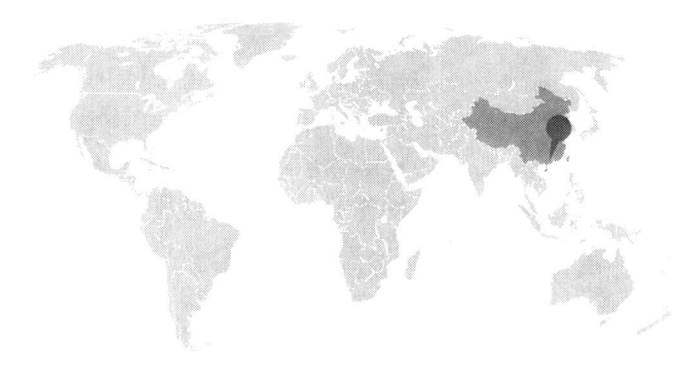

THIRTEEN

Negotiating for a PTA Project in Guangdong

IN THE MID-'90S, I RETURNED TO CHINA AS PART OF A TEAM to negotiate the commercial and financial terms for a US $400 million Purified Terephthalic Acid (PTA) plant in Zhuhai in Guangdong Province in southern China. PTA was a petroleum derivative used as the feedstock to make polyester fiber, plastic bottle resin, and other useful products. Zhuhai was located just across the border from what was then the Portuguese colony of Macau. One would go through passport exit formalities on the Macau side and then trek across about a half-mile stretch of "no-man's-land" before reaching the PRC frontier and its border agents. It was always great fun in the heat when toting luggage for a several-day stay.

Zhuhai, an emerging Chinese metropolis, had infamously constructed a large airport on the theory that "if we build it, they (the air traffic) will come." It was a terrible miscalculation since Hong Kong was just completing its newly expanded facility on Chek Lap Kok, a tiny piece of land adjacent to the larger Lantau Island. The Zhuhai airport sat empty with construction cranes appearing to be rusted in place.

I was the lead negotiator on the financing for the Zhuhai plant, including a debt facility that we were seeking from the People's Bank of China,

China's central bank. It would reportedly be the largest loan ever granted by the state lender, and we had some apprehension whether or not we would succeed in closing the transaction. I discovered the importance of the Chinese government designating a sector of the economy as "high priority" and the State Planning Committee accordingly including the project in its five-year plan. The project ticked the box on both these scores.

My opposite number, Mr. Chu, was an uncompromising hardline negotiator who was predisposed to waiting us out in the negotiations. He knew his American counterpart would be anxious to meet established corporate deadlines for the project. There was little incentive, therefore, for him to take the personal risk of reaching an agreement prematurely.

I realized that I was going to need to earn Chu's trust and build a closer working relationship. Early on in the discussions, I invited him to dinner, suggesting that he pick the restaurant since he would know the range of possibilities far better than I. I figured it would also let him feel that he was still "in control," which seemed to be a preoccupation for him.

On the day of the dinner, we had had a particularly difficult negotiating session where neither side gave an inch, and in fact, we seemed to be regressing a bit. I decided later that I must have offended him somehow since he took me to an eatery that seemingly offered only "discarded body parts" as menu options. These were things like calves' kidneys, pig snouts, and the like. I suspected that my "favorite" delicacy from Azerbaijan, aorta stew, must be on the menu somewhere if only I could decipher the Chinese pictograms. Of course, I was totally at Chu's mercy as I neither spoke nor read any Chinese. Sure enough, a robust serving of discarded body parts was placed in front of me as he beamed proudly, knowing he had bested me on this occasion. For all I knew, maybe he really liked these things and was seeking to share his favorites with me. I suspected not, however, as his plate had items which looked marginally more appetizing. I was confident that my original analysis, that he was chastising me for being too uncooperative and stubborn at the negotiating table, was the right one.

I managed to build a reasonable rapport with Chu, and eventually we successfully concluded the loan agreement. There were other dinners, but our local office picked the venues, which suited me just fine. On one such occasion, Chu noticed I was enjoying the beef dish next to me.

"You like da beef? It is American recipe." He once again was smiling broadly. My first thought was that there was a hidden ingredient that would later ravage me, but he quickly added, "Beef in Coca-Cola." The dish was exceedingly tender and tasty. I knew that Coke was acidic and hence could be used for a variety of household purposes. It would chemically break down various things, but I had never considered using it as a marinade.

"Yes, very good and very creative!" I responded.

I pointed to another dish inquisitively. I knew that a) it might be considered rude to inquire, and b) I might be better off being ignorant. Still, perhaps ill-advisedly, I plowed forward.

"So, Mr. Chu, in Texas we call these fajitas," I said, referring to the tortilla-like items that contained a light-colored meat. "What is the meat inside?"

"I not sure of English word. Maybe it called 'white dove,'" he replied. It was very good, but I knew that upon my return home my daughter Kathy would berate me to no end for consuming the graceful "bird of peace."

In Guangdong Province, it was not unusual to see birds and pigs in close proximity, a recipe, I am told, for illnesses to cross over from birds to pigs and thence to humans. This part of the world seemed to be the incubator for various diseases, including the infamous SARS (severe acute respiratory syndrome), which began in late 2002 in this very province,[1] and various strains of avian flu. In addition, the combination of tropical heat, limited refrigeration, and less-than-rigorous sanitary standards meant that bacterial infections were not uncommon. Indeed, I contracted food poisoning in Zhuhai on one visit. I managed to make it back to the US before falling quite ill and requiring rehydration.

◎ NEGOTIATIONS PRINCIPLES

In every culture, there is no substitute for building a bond of trust between negotiating parties. It helps to increase understanding and can be disarming as both teams show their human side. It is essential to be aware of a couple of huge misunderstandings on this point, however. First, building rapport does not entail gratuitously volunteering concessions as a way to ingratiate oneself with the opposition. In fact, various cultures, including those in China, Russia, Vietnam, and the Middle East, respect hard-nosed and well-informed negotiators who have done their homework and are creative in trying to close negotiating gaps. Indeed, they do not respect those who seem eager to "give away the farm." While compromise and flexibility are expected, it is not anticipated that a negotiator would jump to the end game prematurely without fully understanding the dynamics and components of the discussions.

Second, one has to respect the culture. In the Orient, for example, face-saving is a huge factor. In seeking solutions or offering criticisms, considerable care must be exercised to avoid insulting the counterparty. The Chinese side may struggle with the idea of reversing course, for example, on a draft agreement provision that they had said heretofore was "not negotiable." It can help immeasurably for the Western party to fashion an opportunity for the Chinese side to change its position with minimal embarrassment.

It is also essential to gain an understanding of the other party's key motivators. This is usually done through inference and research, rather than via frontal assault, especially in China. That said, getting to know a negotiator personally sometimes leads to a better understanding of his or her goals and constraints in the discussions. Sometimes the apparent motivators in a negotiation are just superficial, perhaps even a mask obscuring the true underlying drivers.

In the Far East, I found that negotiators tended to have rather odd, indirect, and subtle ways of saying no. They were far less straightforward than businesspeople from cultures more similar to our own. This reality made "reading them" even more difficult than usual. When they said no, might they actually mean yes under certain conditions? I also found that if the stars were properly aligned (meaning the project had internal PRC approvals and had been assigned a high priority), then the wheels of motion could actually turn quite quickly. If the reverse were true, the Chinese counterparty might sound interested to avoid creating offense, but real demonstrable progress was quite unlikely.

As the Chinese economy continued to grow, its consumption of petroleum quickly outstripped currently identified sources, both indigenous and foreign. The need to find additional reserves around the world became a top imperative for the PRC. This prompted the Chinese state energy companies, such as CNOOC and the Chinese National Petroleum Company (CNPC), to pay large sums to acquire exploration and production rights, sometimes greatly exceeding what other international companies could economically justify.

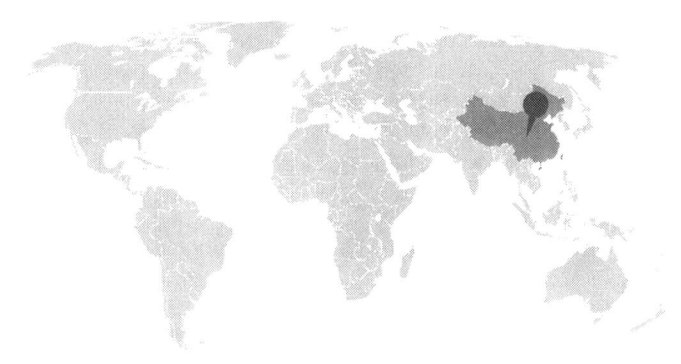

FOURTEEN

Surrounded

ONE OF MY MORE RECENT NEGOTIATIONS WITH THE Chinese was with PetroChina (a division of CNPC) in central China. At that time I was employed by Burlington Resources (since then acquired by ConocoPhillips), and we were pursuing the rights to a "tight gas" project. Burlington had a number of tight gas wells in the US and was adept at enhancing the production rates from these gas reservoirs through the use of hydraulic fracturing, or "fracking." The block was a sizeable one located north of the city of Chengdu, a metropolis that a business colleague once aptly described as "the largest city you have never heard of."

I went along on one visit to the field, located in the verdant hills north of Chengdu, a city perhaps more noteworthy for its panda bear exhibitions than for its oil and gas production. The difficulty with onshore projects in the PRC was that the Chinese government, acting through various state entities, had you surrounded. They set the prices that could be charged for gas sales. They were the sole purchaser of the gas. Burlington was required to use their drilling rigs. PetroChina was our partner, but the Chinese also controlled pipeline access and the associated transportation tariffs. As a consequence, Chinese entities had perfect information on all facets of the project and knew precisely how to extract as much value as possible, while still leaving a modest incentive for Burlington to proceed with the project. Add to this scenario considerable technical uncertainty about the performance of the wells following the fracking (results varied substantially from well to well), and we were left with a rather risky affair, the economic outcome of which was largely outside our control.

PetroChina's local representatives insisted upon the usual banquet and imbibing of Bai Jiu. One of our delegation, an explorationist named Don, had recently undergone heart bypass surgery in Houston, and the doctors had specifically instructed him to avoid any alcohol. As the toasting began, we noticed that Don was downing the liquor in a noble but ill-advised attempt to support our team effort and to avoid insulting our hosts. This occasion was particularly bad since we were outnumbered by the PetroChina personnel. The Chinese would get up and come to where we were seated and toast each of us in turn. The honoree was expected to down the drink entirely, while the other folks at the table did not participate, choosing only to observe. Several colleagues and I rushed over to Don and urged him to stop drinking. Any corporate benefit was simply not worth the personal risk. We explained the situation to PetroChina, who seemed to understand and sympathize. I slept miserably that night, partly worrying about Don's health and partly choking up the Bai Jiu, which burns every bit as much the second time around! Don came through the ordeal, ostensibly no worse for wear.

In Chengdu and the surrounding area, the ritual for the uninitiated was the Sichuan hot pot dinner. I had been forewarned by colleagues to avoid that experience at all cost since the spiciness of the meal was reportedly "off the scale." Hot pot dinners were popular throughout China, but even residents from other parts of the country avoided Sichuan hot pot for good reason. I have since tried (and enjoyed) hot pot in Shanghai, but I suspect that it was a substantially diluted version.

It was fascinating to see the well locations and to observe firsthand the various physical, technical, and commercial challenges we faced in our exploration and production operations. It was rather hilly terrain in what were the lowest foothills of the Himalayan mountain chain.

Burlington was pursuing another project (coal bed methane) in northern China. In addition to the challenges faced by the Chengdu project, another one surfaced for this venture during a negotiating session in Beijing with partner PetroChina. Although no agreement had yet been signed, PetroChina quietly had begun already to drill independently on the block, perhaps hoping that fact would not be noticed until sometime down the road. This revelation certainly did nothing to build trust between the parties and was reminiscent of another PRC experience—this one in the Bohai (Yellow Sea) where a geologic map had been altered by the Chinese conveniently to remove the dry holes (unproductive wells) that had been drilled. Such fraudulent activity, while certainly not the norm in China, can result quickly in soured relations and a project not going forward, at least not with Western funds and technical support.

After ConocoPhillips took over Burlington's international portfolio, it opted to back away from the coal bed methane project.

⊘ NEGOTIATIONS PRINCIPLES

Conflict of interest is a concept that apparently is not fully understood, or at least not highly valued, in some cultures. When government-controlled entities fill as many roles as they did on the Sichuan tight gas project, some of their objectives will be at odds with ours. Yes, it is true that PetroChina was our partner, but it and other government entities had ways of siphoning off economic rent that would leave us with a very marginal project. In fact, the state's economic benefit outside the project likely dwarfed the potential rent that might flow to PetroChina in its role as project participant. Moreover, as a joint venturer in the upstream, PetroChina would be privy to all of the company's technical analyses and forecasts and knew precisely what price the company would need to earn a minimal rate of return. It could (most likely, would) then pass along this information to the pipeline company and drilling entity with predictable results.

A coal bed methane project (also known as coal seam gas) requires the drilling of a great number of relatively shallow wells (each with anticipated steep production declines) in order to generate enough present value to make the project worthwhile. This aspect would have been challenging enough without the added complication of surreptitious drilling on the block by PetroChina. A CBM (CSG) project more closely resembles a manufacturing operation than it does conventional gas production.

Burlington was not in control of its own destiny with these projects. ConocoPhillips, in my view, made the right choice to look elsewhere for gas reserves.

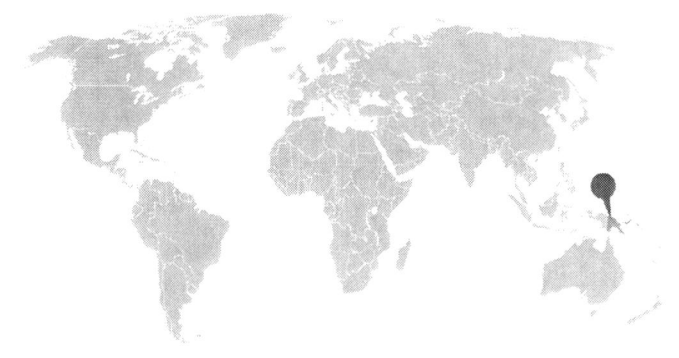

FIFTEEN

To the Ends of the Earth
Following in Rockefeller's Footsteps

A NUMBER OF COUNTRIES IN THE WORLD SHARE THE name "Guinea"—the Republic of Guinea and Equatorial Guinea, both in West Africa, and Papua New Guinea, or "PNG" as it is commonly abbreviated, in the South Pacific. PNG, which lies about 150 kilometers (90 miles) north of the Australian continent across the Torres Strait, shares the island of New Guinea with the Indonesian province of Papua (formerly Irian Jaya). The nation, which also includes over six hundred smaller islands, is rich in minerals and hydrocarbons. Transportation and internal communications in the country have been hampered over the centuries by particularly steep and rugged mountains and by impenetrable rain forests. As a consequence, parts of PNG were so isolated that it had over 750 separate languages because pockets of indigenous people rarely had contact with other tribes or the outside world.[1] PNG was governed for a time by Australia but was granted full independence peacefully in 1975. Since then, Australia and PNG have maintained a close working relationship.

Because of its remoteness, exotic rain forests, and fascinating people, PNG has long been a powerful lure for anthropologists and archaeologists, as well as those seeking an unusual, yet rewarding, holiday. One of those attracted to the South Pacific nation years ago was Michael Rockefeller, youngest son of former US Vice President Nelson Rockefeller. After graduating from Harvard in 1960, Michael Rockefeller journeyed to Netherlands New Guinea, as the territory was called then, with an expedition sponsored by the Harvard Peabody Museum of Archaeology and Ethnology to study the Dani tribe in the west of the country.[2] Michael was also fascinated by the Asmat tribe and its art. Accordingly, after the initial expedition concluded, he elected to return to the southwestern part of the country to conduct further study.[3] Rockefeller was quoted as saying, "It's the desire to do something adventurous at a time when frontiers, in the real sense of the word, are disappearing."[4]

On November 17, 1961, Rockefeller and Dutch anthropologist René Wassing were crossing the mouth of a river on the southwestern coast of Netherlands New Guinea in a heavily laden double-hulled catamaran when several waves from the Arafura Sea flooded their outboard motor.[5] Thereafter, the craft floated out to sea and overturned in the rough seas. Wassing wisely elected to stay with the boat. Meanwhile, Rockefeller, proclaiming, "I think I can make it [to shore]," swam off.[6, 7] The next day a rescue team successfully reached Wassing, who survived the incident. However, Rockefeller could not be found despite intensive search efforts.[8] His disappearance is one of the more intriguing mysteries of the twentieth century. Various theories have been advanced ranging from his drowning, to his being devoured by sharks or saltwater crocodiles (which can range up to 20 feet in length), to his making it ashore only to be consumed by cannibals.[9] Cannibalism was still practiced in parts of Netherlands New Guinea at the time. Rockefeller's body was never found, and he was declared legally dead in 1964.[10]

In his 2014 book, *Savage Harvest*, investigative travel journalist Carl Hoffman has built the case that Rockefeller was killed and eaten by the Asmat tribe. He asserts that the motivation was revenge for an earlier massacre of several tribe members by a "trigger-happy Dutch military officer."[11, 12]

History records that not long after Rockefeller vanished, scientists from Kennecott Corporation in the mid-1960s were flying over a particularly remote section on the west side of the country when they spotted a gap in the otherwise virtually impenetrable rain forest. They decided to set down their helicopter. Indigenous people poured out of the bush, believing the Kennecott crew to be supernatural beings. In the Western Province of what is now PNG, the locals were unaware of the existence of the wheel until the 1930s. Accordingly, one can only imagine their shock at seeing a helicopter landing in their midst. The encounter was peaceful, and the chopper returned to its base without incident.

In the 1980s, the population of PNG's Western Province had a life expectancy of just over thirty years. Their diet consisted largely of taro, a root crop, as there was little game to hunt. Disease was rampant, including a particularly lethal form of malaria, elephantiasis (*lymphatic filariasis*), and dengue fever, to name a few. Adding to the ambiance, this area of PNG also boasted one of the largest, most diverse, and colorful collections of spiders in the world.

The scientists postulated that the reason for the absence of vegetation where the Kennecott crew's helicopter had alighted might well be the presence of a large ore body. Their instincts and professional judgment were correct. Today it is the site of the Ok Tedi gold and copper mining project. The ore body included a gold cap, which was nearly pure gold, overlying rock that contained a mixture of gold and copper. In the early days of the mining operation, gold was flown out in the form of ingots aboard heavily guarded aircraft.

A consortium of companies, having acquired the Ok Tedi project from Kennecott, was developing it in the 1980s when I became involved. Amoco had spun off Cyprus Minerals, which contained all of its other hard minerals businesses, to shareholders in 1985.

In addition to Amoco, participants in the Ok Tedi venture included the Australian company BHP, the State of Papua New Guinea, and a triumvirate of German firms—Metallgesellschaft AG, Degussa AG, and DEG—Deutsche Finanzierungsgesellschaft für Beteiligungen in Entwicklungsländern GmbH. Nobody in the project—certainly no participants whose native tongue was English—ever referred to DEG by its full name for obvious reasons!

In 1986, a small group was established in Amoco's Houston office to manage its Ok Tedi shareholding. I led this team and reported directly to Mr. Blanton, whom I have spoken of previously. There were really two pressing objectives at the time—1) take two large syndicated loans nonrecourse by meeting the requirements set forth in the loan documents, and 2) quietly search for a purchaser of the project. At the time, Amoco Corporation stood behind the two large loans but wanted to convert them to nonrecourse status. "Nonrecourse" meant that the lenders would be dependent for repayment solely upon the cash flows from the project itself. That would mean that the lenders would not have access to the general financial resources of the company or its affiliates. In addition, nonrecourse loans generally attract more favorable "off-book" accounting treatment. The loan documents were highly complex and detailed and would require periods of intense work activity that could only be accomplished practically at the mine site.

My first visit to the mine site in Tabubil, Western Province, PNG, occurred within days of the July 13, 1986, birth of my second child, Stephen. I was not enthused about leaving Houston at a time when most fathers would be bonding with their newborns, but I knew duty called and I really had no choice.

PNG was not the easiest of destinations to reach. Probably the simplest routing was to fly to Sydney, Australia, change planes, and then travel north to the PNG capital of Port Moresby. Alternatively, I discovered later that one could fly from Honolulu to Cairns in northern

Queensland and then take a company charter to the mine. I opted for the latter route on subsequent trips, as Cairns proved to be a more enjoyable and safer stopover than Port Moresby. Cairns, situated on the tropical north Queensland coast, is a popular tourist destination for those seeking the incomparable experience of scuba diving or snorkeling on the Great Barrier Reef.

I found it humorous that the PNG government would insist upon the spraying of the airline cabin with insecticide prior to landing in PNG, apparently as a "tit-for-tat" response to an identical demand from the Aussies that any flights arriving from the island nation be similarly sprayed. I could not help but wonder if the supposedly harmless spray was as benign to humans as the flight attendants claimed. If it was so innocuous, how was it so lethal to mosquitoes? As the crew suggested, I would cover my face with an airplane pillow and wait until the spray had sufficiently diffused (or I had passed out!).

At the time, Port Moresby was a notoriously dangerous place with an elevated crime rate and substantial unemployment. We stayed at the Travelodge just outside of town, dined on a suitable "last supper," and then left for the mine on a company charter plane the next morning. The Travelodge offered a delectable lobster thermidor, which made the overnight stop worthwhile. While the South Pacific lobster tasted a bit mealier than its North Atlantic cousin, it still was delightful drowned in a mushroom cream and cheese sauce. The layover also helped us recover from jet lag.

Partway through my evening meal at the hotel, dining with business colleagues, I lost a portion of a molar, exposing a razor-sharp amalgam filling. This would normally not be an earth-shaking event—in fact, it would not even bear mention—were it not for the fact that we were to depart the next morning for a week's stay at the mining camp in Tabubil, which I pictured lacked modern dentistry. Virtually every time I spoke, my tongue was lacerated by the razor-sharp metal edges of the exposed filling. I knew I could not wait until my return to the US to deal with this situation. I was assured by a more experienced traveler to the mine site that Ok Tedi had a reputable and fully capable resident dentist.

The roughly two-hour flight to Tabubil from Port Moresby in a small two-engine aircraft was routinely bumpy. Warm, moist air from the sea would blow up against the precipitous and majestic mountains, generating convection and creating storm clouds. The mine received over 310 inches of rain per year, making it one of the wettest places on the planet. Although the day had dawned hot and sunny in Port Morseby, the sky filled ominously with soaring thunderheads as we approached the mine. The plane increasingly rocked, pitched, and rolled.

The "airport" in Tabubil consisted of a single dirt and gravel landing strip with little room for error. Undershoot the runway and one would crash into the jungle canopy; overshoot it and a towering, nearly vertical, escarpment loomed dead ahead. In fact, the mountain face already displayed a large pock mark, a solemn reminder of an earlier air calamity. We touched down, bounced around a bit, but eventually came to a successful stop. Several vehicles were parked near the landing strip, prepared to take us to our rather basic accommodations in prefabricated cabins.

I was greeted by the news that the dentist was on leave, but "the veterinarian would meet me at the dental clinic." My mouth sore and bleeding, I realized I needed to rely upon the vet. He was a nice enough chap, but I could only imagine that his experience in treating human dental needs was somewhere between limited and nonexistent. I figured he would fashion some sort of soft covering to overlay the filling as a temporary measure until I could get back to civilization. I was wrong. He mixed up a silvery compound in a dish on the counter next to me and otherwise prepared himself for the procedure at hand.

"I don't really know where the right tools are," he admitted as he stuck a washcloth in my mouth to keep my tongue safely at bay (and perhaps obviate any complaining from his patient). I could feel myself nearly gagging on the rag as he began the procedure. He tapped the material into place and smoothed it with a sanding device.

"Well, that should solve the immediate problem!" he proudly declared. I ran my tongue over the area. I had to admit that, by comparison, it felt as smooth as silk. I thanked the vet, who probably had worked most recently on the mouths of German shepherds, and returned to join my colleagues.

Back in Houston, two weeks later, I would visit my dentist. I recounted my story in abbreviated form. Dr. McConney peered into my mouth and erupted in raucous laughter.

"Come in here, Sheryl. Have a look at this!"

Suddenly, the entire dental office was agape, staring at my inelegant and amorphous filling that, of course, required painful drilling to remove and replace. It was truly a cacophonous chorus of uncontrolled laughter. I was rather embarrassed, to say the least. I knew that the staff would use every opportunity to tease me about the infamous filling during my return visits to the dental office.

Meanwhile, back in Tabubil, I settled into my modest surroundings and attended a meeting on the daunting and tedious process that lay before us. The mine had to meet certain financial and operational tests in order for the loans to qualify for nonrecourse status, and the documentation had to be prepared and submitted to the banks for their review and approval. If Ok Tedi failed any one of the tests during a given month, the process could be repeated, but it meant completely starting over. In other words, passing a threshold test in one accounting period had no carryover to the next; it needed to be repeated successfully.

That evening, a social gathering was planned for us directly beneath the CFO's house, which was constructed on stilts. It was scheduled for 7:00 p.m., well after dark, which did not seem like a great idea since every self-respecting mosquito in the Western Province would likely turn up uninvited for the event. The temperature had dropped considerably from daytime levels, but moisture hung thickly in the night air. I doused myself with DEET, wore loose-fitting, long-sleeved clothing, and must have resembled an American baseball third-base coach touching, in turn, various parts of my body in a relentless effort to brush away any would-be insect assassins. I was taking antimalarial medication, but one could not be too careful. The local workers took distinct pleasure in telling us that the variety of malaria they had in western PNG would manifest itself initially as a pain near the back of the head. One would be dead within hours thereafter, they claimed. I suspected that they were trying to torment their visitors, but I did not want to take any chances.

The Aussies also particularly enjoyed reminding their North American visitors that there were cases of local cannibalism as recently as the late '60s/early '70s. "Once you have acquired a taste for people, mate, you never really lose that craving," they would delight in proclaiming. I recall the face of a senior Amoco executive, in town for a quick tour of the mining facilities, growing ghostly pale after hearing that story. Of course, I added to his misery by reminding him cruelly that I was all "skin and bones" compared to him. Accordingly, I represented a mere appetizing morsel, whereas I pointed out that in contrast he more closely resembled a full meal.

The food, both at the welcoming party and at the dining hall, was pretty innocuous, though somewhat repetitious and rather devoid of seasoning or spice. I figured the fare had to be decent to prevent an all-out revolt by the imported workers, many of whom were from the Philippines and Indonesia. Morale is important at such a remote outpost, and food is an important part of that equation. I brought a variety of snacks to supplement the meals or in case I should miss one. After all, there were no restaurants within hundreds of miles of the mine and no practical way to reach them anyway!

The next day the work began in earnest. It was tedious labor, which entailed gathering financial facts and figures that would later be organized, analyzed, and then presented to the banks. The discussions with the banks were quite detailed and complex. I have not provided further elaboration as the specifics are not essential to our story.

It was interesting to work in these conditions. Several female secretaries, who lived locally, tapped away at their keyboards, capturing and recording our data. One had a strange pet positioned on her shoulder, watching intently as she typed. It had large, round, yellow eyes and medium-length gray fur. I was told it was a local marsupial, some sort of cuscus. Far from a cuddly pet, it looked more like a menacing creature from a horror movie with its bulging yellowish/orange eyes! If this animal were not unusual and distracting enough, the secretary also sported a Cross pen, complete with the Ok Tedi corporate logo, inserted horizontally through her nose. While I had seen tribespeople with objects protruding through their nostrils, I had not witnessed a corporate pen used in this fashion. Trying not to stare, I recall grimacing a bit as I considered the pain that must have accompanied the original insertion.

Late that afternoon back in our accommodations, our lawyer from Chicago spied my cache of snacks and proclaimed, "I will not be eating Spam, regardless of how bad the food gets here." By the end of the week, however, he seemed to have reconsidered his position and experienced an epiphany. I spotted him eyeing my Spam longingly, though his pride must have kept him from inquiring as to its availability and price! I was spoiling for a negotiation!

It rained heavily during most of our visit, punctuated with the odd period of sunshine. When the sun did emerge, the outside temperature would soar, bubbling up the steamy atmosphere and prompting it to precipitate once again. The rays of sunshine sparkled off the myriad spiderwebs that were still coated with the liquid residue from the recent rains. As someone who is quite tall, I graciously (but unintentionally) cleared the path for others as my head snared each lofty sticky web at my elevation.

A festival to commemorate a local holiday was scheduled on the

Friday before we left the camp. Company employees, visitors, and local tribes were invited. Many of the latter came out of the woods, their faces painted, carrying bows and arrows and wearing penis gourds and loin cloths or grass skirts. I suspected that they may have dressed up slightly for the occasion, but I was assured that their customary attire was similar—just less colorful. They enjoyed being photographed with me since I guess they had never seen a human that tall. Diminutive in physical stature, the tribe members were nearly half my height.

Young with local tribe representatives at Tabubil festival

Relations between the indigenous tribes and the local employees seemed to be quite good. Only once do I recall a period of labor strife that escalated to the point of violence.

In such a remote location, stories abounded, of course, many of which I found rather interesting. One story had it that early in the life of the mining project a barge carrying cyanide pellets went aground during a dry period when the river's flow was minimal. Cyanide was used in the mine-leaching process to recover the highly sought-after gold. Some of the cyanide had gone missing, which was obviously a source of consternation and serious concern. It was reported that one of the tribes was later spotted "fishing" with it, throwing the pellets in the river, watching the dead fish rise to the surface, and then retrieving them from the water. We were aghast to hear this account, both from an environmental perspective and the thought that they intended to cook and consume their catch.

The companies were quite concerned about the ecology of the area, running frequent tests to ensure that the water in the Fly River was not being contaminated. The companies sought to ensure that mine tailings or other pollutants not flow downstream and threaten either life along the river or at the Great Barrier Reef, which lay just beyond the Fly River delta. The reef was, of course, an international treasure. Normally, the currents at the delta would carry any pollution westward away from the reef, but nobody wanted to take the chance of a brief weakening of those currents allowing pollutants to make their way through to the sensitive reef environment.

The project faced a number of challenges, not the least of which was operating in such a remote and expensive location. At the outset, it was planned that equipment would be floated up the Fly River. However, a hundred-year drought materialized, meaning that materials had to be delivered by helicopter to the mine site, which translated into extensive delays and higher costs. In addition, metal prices were not particularly high during the early years. Ok Tedi commenced a hedging program to ensure it would have the cash it needed to service its debt and meet its operational requirements. The project paid its first dividend to shareholders on June 28, 1991.

Some of the companies who participated in the Ok Tedi project discovered ways to profit from the project aside from receiving dividends. Some would charge a fee to market the company's production, while others would supply various goods and services.

> ## ⊙ NEGOTIATIONS PRINCIPLES
>
> Ok Tedi relied upon a cost-plus contract, a fairly common arrangement, for construction of its facilities. One weakness of such an agreement, however, is that it does not incentivize a contractor to minimize its costs. In fact, the higher the costs, the greater the base to which the profit percentage is applied. Let us assume that a contractor has a 10 percent profit component. If a facility could be built for $100 million but the contractor did it for $110 million, the contractor would earn an additional $1 million in the process.
>
> The company I worked for understood conflicts of interest and studiously avoided them. To sell goods or services to a project in which one is a shareholder struck me as a clear conflict of interest unless it could be clearly demonstrated that the joint venture was charged fair market value. To avoid this situation, and the friction among the parties that would inevitably result, joint ventures should agree upon a mechanism such as an arms-length bidding process.

The morning we were scheduled to leave the mine site dawned foggy with rain. We were informed that the visibility, in fact, was insufficient for the corporate plane to land. It was to fly in from Cairns. Instead, we were instructed to load ourselves and our gear onto a bus and take the winding road down to the Fly River town of Kiunga about 100 kilometers (60 miles) away. The weather would likely be better at the lower elevation, and we could rendezvous with the plane there. Realistically, there were only two ways out of this part of PNG—flying (from either Tabubil or Kiunga) or floating down the Fly River to the sea coast. Recalling the Rockefeller misadventure, I was not keen on the river option. Because of the rugged terrain, there was (and is) no comprehensive highway network connecting various parts of the country.

The bus ride to Kiunga was not a pleasant journey. The dirt road was bumpy and the windows were caked with mud from the unrelenting rains after the first mile or so. A number of the passengers suffered from motion sickness, at least in part because of the restricted visibility. Still, we were headed home and that kept us focused. I looked forward to reuniting with my wife, daughter, and baby boy, Steve.

Meanwhile, my team's other major objective was to search for potential buyers for our stakeholding in the mine. With the prior disposition of its other hard mineral businesses, Amoco no longer considered the Ok Tedi project to be a core investment, and it was absorbing a great deal of senior management time and attention. Accordingly, Mr. Blanton instructed my team, with the help of our attorney Sam, to scour the planet for potential buyers.

We prepared a glossy marketing brochure describing the project (yes, one with a solid *gold*-colored cover, of course) and sent inquiries to a wide range of companies. Based upon the responses and our research, we narrowed the field to three potential purchasers—one in the UK and two in Australia. We then conducted a "road tour," giving a presentation and answering questions from the target companies. We entered into negotiations with one of these firms and succeeded in obtaining the approval of their mineral division CEO. Mr. Blanton was ecstatic and declared that the company would "hold a ticker tape parade in my honor." However, exuberance quickly succumbed to dejection when I received a call several days later informing me that the board of the target company, which had powerful external directors, had vetoed the sale. Normally, one would assume that the board would be more of a formality than a major hurdle, but such was not the case in this instance. A couple years later, the mine was ultimately sold on similar terms to partner BHP, but the ticker tape parade never materialized.

As a postscript to the PNG stories, I cannot resist the temptation to relate the account of one of our attorneys who, on a subsequent trip, was to meet up with the rest of the delegation in Cairns prior to our onward corporate flight into Tabubil. Most experienced international travelers know to carry a change of business attire in their carry-on luggage, and Charlene was probably no exception. However, on this occasion she chose to board the flight in Chicago in hot pants (fashion craze of the 1980s) and had checked in as baggage everything else but her attaché case, which contained the documents and business papers that were the focus of the meeting in Tabubil.

The reader has no doubt already leapt to the correct conclusion that Charlene arrived in Cairns with only the shirt on her back and the hot pants on her derrière. She was panicked, and the head of our delegation promptly informed her that he would not permit her to travel on to PNG without proper business attire. It was a Sunday and most stores in Cairns were closed; the onward flight was to depart bright and early on Monday morning. She managed to scrounge up some clothes locally that, while not typically conservative attorney attire, would not result in her immediate arrest upon setting foot in PNG. I must say that she looked more like an attorney on holiday at the beach.

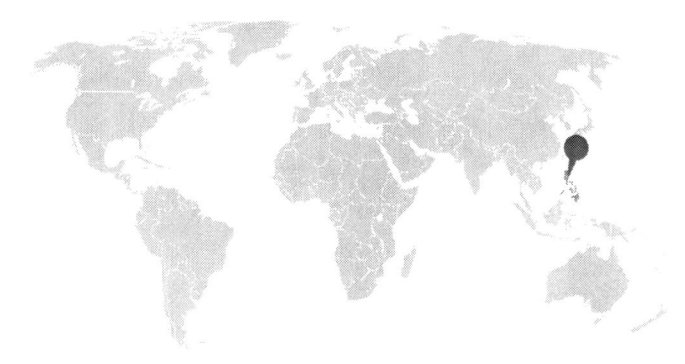

SIXTEEN

Concluding a Deal amid Philippine Chaos

THE PHILIPPINES AND THE US HAVE ALWAYS HAD A LOVE/ hate relationship. The US governed the Philippines from the time of the Spanish-American War in 1898 until after World War II when the nation was granted full independence. The Clark and Subic Bay bases were among the most strategic American military facilities in the Pacific. Having tired of the US presence and feeling that they no longer needed the Americans to protect them from regional adversaries, the Philippines ordered the bases closed in the early 1990s.

The US government had supported the regime of Ferdinand Marcos for years, but in 1986, change was afoot in the island nation. Corazon Aquino, wife of assassinated senator Benigno Aquino, an outspoken critic of the Marcos regime, was running as a democratic alternative to Mr. Marcos's authoritarian rule. Allegations were rampant that the Marcos regime was corrupt and siphoning off aid from the US and diverting it for personal use.[1] Imelda Marcos, the president's wife, was legendary for her extravagant lifestyle. Her excesses included accumulating over one thousand pairs of designer shoes; original paintings by the likes of Van Gogh, Cezanne, Rembrandt, Rafael, and Michelangelo; and silver tableware, gold necklaces, and diamond tiaras.[2] As of 2013, only about 25 percent of the US $10 billion in assets that the Marcos family amassed, allegedly through theft, had been recovered. These excesses did not play well with the Philippine population, many of whom were desperately poor and were convinced the Marcos family was indeed stealing from them.

A snap election called by President Marcos resulted in him being reelected. However, the popular belief was that Mr. Marcos had stolen the election from the popular US-educated Aquino. Street protests grew into a popular uprising, which became known as "People Power," as the pressure ratcheted up on Mr. Marcos to step down. When the military withdrew its support, the president eventually fled with his family to Hawaii.[3]

It was against this backdrop that in late 1985 I was asked to negotiate the sale of Amoco Philippines and to close the transaction before any serious trouble—like a coup d'état, anarchy, or a civil war—erupted. Amoco's only production was from a small field called Cadlao in the South China Sea off of Palawan Island. The field was developed using one of the first ever subsea well completions, which produced into a Floating Production Storage and Offloading (FPSO) vessel. An FPSO was a converted tanker that would store the oil until another vessel pulled alongside, emptied the FPSO of its contents, and sailed off with its precious cargo.

Despite being a technical success, the Cadlao field was not of sufficient scale to merit the disproportionate time required of senior management to deal with what seemed like an endless stream of issues. The field was rapidly approaching its economic limit (defined as the point when after-tax net production revenue approximated the costs of continued operations). Amoco's management felt that the economic limit date could be extended out in time if the field were sold to a small independent oil company that could operate the field in a low-cost, bare-bones manner.

Before selling the asset, however, Amoco opted first to try to renegotiate a lower bareboat charter (lease) rate for the FPSO. For this purpose, an Amoco delegation (myself included) met with FPSO owner SBM, a subsidiary of Dutch company Heerema, both in Houston and at its offices in Monte Carlo. Unfortunately, we were unable to secure a low enough lease rate to merit hanging onto the asset.

Negotiators rarely are dispatched to glamorous destinations like Monaco, and my one and only business trip to Monte Carlo occurred while I was playing the role of economist. As the old saying goes, "Oil is rarely found in nice places." Based upon my observations and extensive travel, this adage seemed to be spot on.

I put up little resistance—actually none—when approached to make the journey. We stayed at the Hotel de Paris, immediately adjacent to the main casino in central Monte Carlo. The Hotel de Paris was an exquisitely appointed and grand hotel, which had previously hosted famous entertainers and heads of state, including Cary Grant, Michael Jackson, and Nelson Mandela.[4] I discovered that Monaco had a centralized wine cellar, one of the largest and oldest in the world, which was connected by a series of tunnels to several of its restaurants. The lawyer on this business trip, Jerry, was a wine aficionado and connoisseur, so I felt compelled during my stay to sample a few of the French wines he recommended.

A portion of the team, myself included, traveled on to Manila to meet with the Amoco personnel based there. Our objective was to begin to assemble the information that we would need to market the Philippines affiliate. We needed to keep our intentions to ourselves for the time being. The local office ran very efficiently with one of the better staffs we had anywhere in the world.

Back in Houston, we succeeded in identifying a small independent oil and gas company, Alcorn Energy, which already had some level of activity in the Philippines and which was interested in adding the Cadlao field to its portfolio. Alcorn knew we were not desperate to dispose of the asset, so we were able to negotiate a fair price. We still had to sell our last load of oil in the FPSO, the proceeds of which we would be entitled to retain. We also had to draft and finalize the various closing documents. Conditions on the ground in Manila were continuing to deteriorate, so my other attorney Richard and I were under some pressure to complete the sale without delay. It appeared that the best place to accomplish these objectives was back in Manila.

By this time, I had gotten to know our Philippines staff. While the purchaser had agreed to retain at least some of the staff, the others would likely lose their jobs. I felt particularly bad about the chief accountant whose father, Manila's chief of police, had recently been assassinated. I certainly did not want to contribute any further to her family's sorrow and hardship. However, it appeared likely that she would be retained. Then there was Amoco's resident manager in the Philippines. He had concluded that the real nature of our visit was to gather data for a planned divestiture and saw that action as the company's covert tactic to force him into early retirement. Actually, while he was correct as to the first point, he was entirely off the mark on the second.

Manila was a city of contrasts. The commercial district, Makati, was built around the layout of the old airstrip from World War II. That facility had three crossing runways, forming a triangle, and the triangular nature of Makati remains intact to the present day. Not far from the business district, on the way to the international airport, one passed the barrios in which many impoverished Filipinos lived in very difficult conditions.

The Philippine capital was also a vibrant city with five-star international hotels, modern shopping, and a notorious red-light district around Mabini Street, which had sprung up to service the US military men who were present in large numbers prior to the base closures. The Filipinos have sometimes been called the "musicians of the Pacific" and with good reason. Many of them have produced some great music and possess an amazing talent to imitate famous foreign singers with incredible accuracy. I would sit in the bar at the Mandarin Hotel after dinner and listen to a local artist singing "Say You, Say Me." As I sipped my beer, I closed my eyes and listened. I could barely discern any difference in the sound from Lionel Richie himself. I could not detect any Asian accent whatsoever.

Concluding a Deal amid Philippine Chaos

During the day, Richard and I would work at a long wooden table in the large conference room at the office facility in Makati. Sporadically, we would hear gunshots ring out during periods of rowdiness in the square—or rather triangle—below. Stories were circulating that anyone not in possession of an airline ticket for a confirmed flight out of Manila was likely condemned to remain in the capital until the power struggle played out and the unrest had subsided. The situation was quite tense, and it was not at all clear that the dispute between Marcos and Aquino would be resolved peacefully.

The crude price had begun to fall worldwide, and Amoco believed that trend would likely continue. We met with the Philippines entity that normally purchased our oil to discuss the price we would receive for our last tanker load of oil. The negotiation included compromises on both sides and additional issues aside from price. However, for the last load of crude oil, we agreed to use the reference price quoted in *Platt's Oilgram* for the prior week. As a consequence, we realized a significantly higher payment than using the reference price for the current week (or, as it turned out, future weeks).

En route back to my hotel from one of my Manila meetings, I hailed a cab. Quickly glancing at the front seat to ensure that the vehicle was equipped with a meter, I hopped in the back. A thick plastic transparent pane separated my seat from the driver in front, a common feature installed to enhance security. It was midafternoon and the relentless sun was baking down, making the backseat unbearably hot. As the cab pulled out into the thick Manila traffic, I knocked on the pane and asked the driver to turn up the air-conditioning.

"Sorry, sir, your rate does not include air-conditioning," came the response from the front seat where the driver was no doubt enjoying a far more comfortable ride. "If you want A/C, you need to pay double rate." Taxi scams are not unusual, though I had not encountered this particular one before. It was either "pay up" or "pass out," in a manner of speaking, so I reluctantly chose the former. This is a good example of one's negotiating position being compromised. Once one is in a locked cab in heavy traffic, there is little option. I guess I could have demanded to be let out, but I doubted that request would have been honored until I had "paid up" in full. It felt like unlawful detention!

I had been asked by a colleague, George, in the Production Department if I would be so kind as to bring back to Houston a pair of leather cowboy boots that he was having handmade in Manila. As I knew I would be fully consumed with negotiations and other meetings while in the Philippines, I did not relish having to perform this additional task. Moreover, I knew the boots would be heavy to transport home. Nevertheless, George was both a friend and in an influential position at the company, so I picked them up the day before returning home. I checked that the box did, in fact, contain boots, and I requested a receipt from the merchant that I could present to the customs agents back in the US.

Other than being extensively and repetitively patted down at the Manila airport, Richard and I experienced no real difficulty departing the Philippines. We were both in possession of valid tickets on United Airlines. Our route of flight took us to Seattle as the first port of entry into the US. As I cleared customs, the agent took more than a passing interest in the cowboy boots for some reason. Perhaps I did not appear to be the "cowboy type." I explained that they were manufactured in Manila and that I had the receipt. He glanced at the receipt, reached down into the boots, and pulled out what appeared to be a thin plastic bag containing white powder.

"What's this?" the inspector asked, his eyes penetrating me like a laser. I could feel my pulse racing as I feared the worst. It had not occurred to me to explore the toe area of the boots. I could see myself mistakenly accused of being a drug trafficker. I knew that ignorance of what was stashed in the boots—even if the agent believed me—would be no excuse. I would undoubtedly be arrested.

"I picked up the boots for a friend who was having them made," I admitted. "I'm afraid that I never looked inside them."

I watched sullenly as the agent unwound the plastic. Mercifully, it was just a piece of thin plastic wrap with a paper tissue buried inside and nothing more. What a relief! Even though this incident occurred well before the days of heightened airport security and warnings never to carry packages for other people, I required nobody's encouragement to vow never again to pick up and transport anyone else's goods.

The Cadlao field sale closed with Alcorn shortly after my return to the US in January 1986. I recall that we had planned to celebrate the closing of the transaction. However, we learned during our meeting that the NASA Shuttle *Challenger* had exploded shortly after launch. Shocked and deeply saddened, we chose to cancel the planned celebration. In one article, which appeared several weeks later, Amoco was lauded for completing the "best-timed deal of the year."

◎ NEGOTIATIONS PRINCIPLES

The sale of Amoco Philippines was an example of a win/win negotiation. Given its lower operating cost structure, Alcorn acquired a quality asset that it could produce for a considerable time. Amoco shed a time-consuming non-core asset, allowing it to refocus its attention on larger-scale projects that would likely produce more after-tax net present value.

Tying the sale of the final tanker load of crude to the oil price from the prior week's quoted price in *Platt's* was an incremental benefit to Amoco, locking in an agreed price before the reference price fell even further (as it did).

The taxi incident with no air-conditioning (unless a usurious price was paid) was yet another example of the futility of negotiations between two parties with markedly differing levels of power.

SEVENTEEN

An Imperfect Union in Buenos Aires

WHEN ANYONE THINKS OF ARGENTINA, IMAGES OF EVA Perón (Evita) and the tango come to mind. Frankly, as a confirmed carnivore, I think first of Argentine beef followed in rapid succession by empanadas and malbec wine. Argentina is a very diverse and enchanting place. Sadly, until recently, I had spent my time exclusively in Buenos Aires, and even then buried away in conference rooms (a fate shared by many business travelers).

From a business standpoint, Argentina was not without its challenges. In the 1980s, it experienced hyperinflation, the Falklands War (known there as the Malvinas War), and periodic political turmoil. Following Argentina's defeat in the Falklands conflict, the ruling military junta collapsed and was replaced by a civilian government. Carlos Menem was elected president, and under his leadership Argentina emerged as a vibrant economy. Oil and gas production was a prime contributor. The prevailing view in the industry was that the country had further exploration potential. Amoco had operated in Argentina for decades, wisely choosing to focus on the long term rather than being put off by continual economic oscillations—alternating periods of boom and bust.

Since its Argentine business was rather mature, the company had been actively reviewing new in-country growth opportunities. Several possibilities had surfaced over time, but either they were not of interest to Amoco or it failed to capture them. When the chance arose to acquire the Southern Cone assets of the Buenos Aires–based energy company Bridas, Amoco pursued them aggressively. The Argentine firm was run by brothers Carlos and Alejandro Bulgheroni, who at times reminded me a bit of J.R. and Bobby Ewing, respectively, of the American soap opera drama *Dallas*. Both Carlos and Alejandro were deeply involved in running the business, but they employed very different management styles.

There were two factions within Amoco at the time—one which favored acquiring the Bridas interests outright and the other which preferred to form a joint venture company. The advantage of the former was that Amoco could assert full control, avoiding the "messiness" of partnering. The benefit of joint venturing was that the Bulgheroni brothers had valuable contacts in Buenos Aires and could potentially open doors to further oil and gas investment opportunities within Argentina. In the end, the joint venture faction carried the day and a 60/40 Amoco/Bridas company, Pan American Energy, was formed. Pan American Energy was among the largest merger transactions consummated in Latin America at the time.

By 1997, I had moved to Chicago to join the Treasury Department, which arranged and managed the company's financings and also oversaw its largest acquisitions and divestitures. For the Bridas transaction, I was a member of the Amoco negotiations team led by the company's Argentina country manager, Mark, an engineer by training. My primary responsibility was to provide advice on the financial elements of the sale and purchase agreement. I also participated actively in developing team positions on various commercial aspects and was in a great position to witness the fascinating negotiations as they unfolded.

Carlos led the Bridas delegation in most of the discussions and proved to be a consummate negotiator. I learned a great deal about the craft from watching him in action. He was particularly adept at "forum shopping." If he did not get the response he was seeking from Mark, he would work his way up the management chain until he found someone more sympathetic to his point of view that also had the power to act. In addition, Carlos was quite effective at feigning anger and issuing implied threats to pressure the Amoco team. He sensed, correctly I believe, that Amoco was highly motivated to conclude the deal.

In order to assess the value of its remaining oil and gas reserves, Amoco had contracted with a well-respected independent engineering firm to conduct an analysis. The estimate came in higher than the company was expecting, allowing Amoco to receive a higher relative valuation of its assets versus those of Bridas.

> ### ⊙ NEGOTIATIONS PRINCIPLES
> Forum shopping can be an effective negotiating tool. It requires the negotiator to know who the key decision-makers are in the other company. If the negotiator has developed some rapport with those executives, he can just pick up the phone and make his case. If the other company allows this process to go on unchecked, however, it can undercut the authority and effectiveness of its own negotiator. To prevent this outcome, the senior manager at a firm subjected to this tactic should simply refer the caller back to the company's lead negotiator. That response also serves another important purpose: It preserves the more senior executive's ability to intervene later to resolve any intractable issues near the end of the negotiation. Often, however, the senior executive cannot resist the temptation to intervene and take credit for removing a roadblock to the discussions moving forward. In the long term, though, such premature intervention can have the perverse effect of prolonging the discussions. It also means that if and when

further impasses arise, an even more senior representative in the corporate hierarchy must swoop in to resolve the issues.

As noted earlier, the "flinch" or other emotional outbursts can also be employed to good effect as long as they are not overused. If someone offers to sell you a car and verbally suggests a figure that is considered too high, you can purposefully show your revulsion by visibly recoiling in disgust (the flinch). The seller then suspects that his asking price is not only too high, but well in excess of an acceptable figure. For his counterproposal, he may then choose to come in with a lesser figure than he had originally planned.

Certainly there is an element of acting involved in negotiations. A negotiator must be able to "sell his story," which requires equal measures of genuineness, forcefulness, and persuasion. Sometimes bluffing is involved, but negotiators need to be able to deal with the consequences should their bluff be called. It is unwise to make a promise or issue a threat if there is no corresponding commitment to follow through with it.

In a similar vein, the negotiator can choose, as Carlos did, to go into a planned and controlled tantrum now and then to demonstrate his or her annoyance. In a competitive environment where the negotiations with a company could abruptly be terminated only to begin anew with the next highest bidder, this technique can be used quite effectively by the host government to get the front-runner to demonstrate some movement. It is best to accompany such outbursts with a restatement of one's position (or perhaps with a modest concession).

Implied threats, while useful on occasion, should not be employed too often. They can have the unintended effect of undermining trust. If, as is frequently the case, the parties would be joint venture partners at the conclusion of the negotiations, it is not beneficial to erode trust. I like to say that "trust is built up over a great deal of time but can be destroyed in an instant." It is no sin to be a "tough" negotiator as long as one is fair.

Amoco's decision to seek an independent assessment of its reserves was a wise one. Even though it had a number of internal experts who could render such an opinion, the opposing side is far more likely to give credence to an informed view put forth by a well-respected, arms-length third-party expert.

Two additional important negotiations topics are 1) choice of law and 2) dispute resolution. Of course, it may be ideal to select the laws of your home jurisdiction. However, the host country may insist upon its laws being applied to interpret an agreement governing a transaction or activities occurring within its borders. An alternative may be to agree upon the laws of a third country, such as the UK, which has a well-established common law tradition. It is generally not advisable to rely upon the courts of the host country, which may tend to side with the local party involved in a dispute.

Arbitration in a favorable (or at least neutral) location can be a good mechanism for dispute resolution. Consider one of the following: The International Arbitration Rules of the International Centre for Dispute Resolution of the American Arbitration Association (AAA), the London Court of International Arbitration (LCIA), the Arbitration Rules of the Singapore International Arbitration Centre (SIAC), the Rules of the Arbitration Institute of the Stockholm Chamber of Commerce (SCC Institute), or the United Nations Commission on International Trade Law (UNCITRAL) Arbitration Rules.[1] Both choice of law and dispute resolution mechanisms at their essence are legal issues. As always, it is important to work in close consultation with your legal team on these and other contractual issues.

Amoco's partner in Pan American Energy, Bridas, was an Argentine firm, suggesting that there might be local sympathies and support for it in any disagreement. Accordingly, it would be particularly important to have a neutral dispute resolution mechanism and venue.

It is also imperative to have a thorough legal review of the suite of agreements prior to execution. In fact, it is my view that counsel should be at your side from the outset as a trusted member of your team. Involvement of tax advisors early on is also important, since their ability to fashion tax-saving structures is greatest near the beginning of discussions. This point is worth emphasizing—so often experts review plans for a deal too late in the process when opportunities to maximize value and/or minimize risk have already been foreclosed upon.

In joint venture situations, it is particularly important to review what behavior the agreement may unintentionally be encouraging. For example, if the interest rate for late payment into the joint account is too low and there are no other more severe penalties for continual late payments, a party can opt to use this mechanism as a source of inexpensive financing. If a company's cost of borrowing is normally, say, 8 percent and the penalty for late payment is only 5 percent, an incentive exists to delay payments. There is even more incentive if the delinquent payer has limited access to funding.

It is also worth investigating your potential coventurer to determine, for example, if they have a litigious tendency, having been involved in a number of prior lawsuits. If there is a pattern of such behavior, it is not unreasonable to assume that lawsuits may result when disputes arise in the future, as they inevitably do, between partners. The vast majority of firms consider lawsuits costly, a distraction from day-to-day operation of a project or business, and a threat to the healthy chemistry and trust between project participants. For others, lawsuits are seen as a powerful lever.

During the Bridas negotiations, we had one memorable meeting in Buenos Aires with Morgan Stanley. It was a morning session to agree upon next steps in the negotiation process, and a buffet breakfast was included. About midway through the meal, the lead Morgan Stanley representative, Howard, announced, "I suppose you wonder why we picked this specific location at the Four Seasons Hotel for our meeting?" There was no immediate response, so he continued, "I thought you might enjoy dining in Madonna's bedroom." After a pregnant pause, followed by laughter, he went on. "Yes, this is the same suite that Madonna used as her bedroom and operating base during the filming of the movie *Evita*." More chuckling ensued. We were all familiar with her role as the former Argentine first lady who was both much revered and much maligned in Buenos Aires. The film, especially when first released, was quite controversial in Argentina.

After the working session concluded, I decided to walk back to my hotel, something I did rather frequently to get some exercise and to clear my mind. The route took me through a large park where I detected a faint voice coming across a loudspeaker. As it grew louder, I realized it was being broadcast in English and that I recognized the speaker. However, I did not immediately place it since it was entirely out of the normal context. It was, in fact, President Bill Clinton delivering a speech to a group of curious onlookers who had gathered in the park. I certainly was not expecting him to turn up in the Argentine capital! The US had remained officially neutral during the Falklands conflict and seemed to have remained on good terms with both Great Britain and Argentina throughout.

The Bridas transaction was concluded in September 1997, making the new entity Pan American Energy the second-largest producer in Argentina behind YPF S.A. Amoco had a 60 percent interest in the new entity, Pan American Energy.[2] As part of the deal, Bridas acquired a minority stake as well in Amoco's Bolivian exploration and production affiliate.

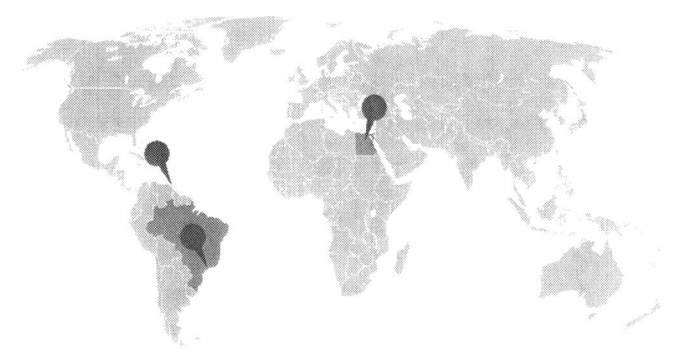

EIGHTEEN

In the Land of the Pharaohs and Beyond

EGYPT HAS LONG BEEN A COUNTRY ASSOCIATED WITH large oil and gas reserves, and Amoco was a key player in unlocking the country's hydrocarbon potential. Legend has it that a young country manager named Jim Vanderbeek drilled what he characterized in his activity reports to senior management as a "water well" but decided to drill it to a depth that would test potential hydrocarbon-bearing zones. The result was a major oil discovery—the El Morgan field in the Gulf of Suez—which led to subsequent other large finds. Not surprisingly, Vanderbeek became an instant hero in Egypt, and his exploration success catapulted his career forward. He was an old-style explorationist who relied upon not only proven scientific methods but also his "nose for oil."

David Work, an explorationist himself, reported directly to Vanderbeek during a portion of his career at Amoco. In a memorial to Vanderbeek, who died in 1993, Work wrote, "Jim had the foresight to look beyond the great political risks of the 1960s and zero in on the promise of Egypt's Gulf of Suez and Western Desert. As a result, Jim and his staff discovered the great Ramadan, El Morgan, and July oil fields."[1]

As is frequently the case in overseas locations, the expatriate community in Cairo was a close-knit group, and most of them lived in rather well-fortified but unobtrusive villas along the tree-lined streets of the Cairo suburb of Maadi. The expat village was just steps from the Nile, the lifeline of Egypt. The exteriors of the local offices of Western energy companies were frequently subdued, and sometimes even unmarked, so as not to attract the attention of protestors or would-be terrorists. For amusement, while they were away from their home countries, the expats would meet at a local watering hole in Maadi one evening each week for a trivia competition. Bragging rights were at stake. Many expats formed teams and took the competition quite seriously.

I recall one evening in April 2000 after BP had acquired Amoco. I was on a business trip to the Egyptian capital, and my colleague Adam and I decided to visit the pyramids just west of the Nile after work. It was a warm evening, but not uncomfortably so. The sun was still shining, though rather faintly. Visibility was somewhat limited, as it normally was. Even on relatively tranquil days, Cairo is victimized frequently by countless small sand particles that are picked up and transported by the winds blowing off the vast Sahara Desert. When one flies into Cairo, the Nile valley appears as a meandering green ribbon, a verdant oasis flanked on both sides by a seemingly endless brown desert.

We had been cautioned by our local office to beware of scams. One notorious ruse pertained to camel rides around the pyramids. A price, which included assistance in boarding the desert animal and a guided tour of the antiquities, was agreed up front. By all accounts, it was great fun and an awe-inspiring experience until the tourist was prepared to dismount from his lofty perch upon the camel, some six or seven feet above the ground. The owner then insisted upon a previously undisclosed (and usually eye-popping) dismounting fee.

Adam and I relied upon a local national from our office, fluent in Arabic, to help us negotiate our tour, and we opted for a horseback ride just in case the merchant was not as reliable and trustworthy as our office representative thought. Dismounting from a horse would not require any assistance from the guide. It was a delightful experience to observe the majesty of the pyramids as the sun sank slowly behind them. I marveled at how our ancient ancestors, lacking modern-day hoisting equipment, could accomplish such an amazing engineering feat. The slabs clearly weighed tons. In some ways, the construction of the pyramids trivialized more recent technological and construction advances.

> ### ⊘ NEGOTIATIONS PRINCIPLES
> The camel scam was a classic case of a disparity in relative negotiating power. When the tourist's feet were still planted firmly on the ground and he was merely considering the features and terms of the ride, he had a reasonably strong negotiating position. After all, he could choose to spend his money elsewhere with another tour operator or not at all. It was reasonable to assume, as most unwary tourists did, that the fee included mounting, dismounting, and the tour itself.
>
> Once the unsuspecting rider was seated proudly atop the camel, his negotiating position was greatly diminished, absent the timely intervention of the authorities who were strangely absent on our visit to the pyramids. In our case, we had the advantage of a) being briefed in advance about the scam and b) having a local national on hand to ensure that we were not "taken for a ride" (at least not in a figurative sense). Although one could seek to pay after the ride was completed, that approach was unlikely to be accepted. The business was also a cash-only and apparently unregulated affair that further limited any recourse.

On another trip to Egypt, a colleague and I were teaching a corporate finance and negotiations course to employees from around the Middle East at the Hilton Ramses Hotel in central Cairo. Located just off Tahrir Square, the facility had marvelous views of the city and the Nile from its upper floors. The broad Nile River appeared majestic yet mysterious as it meandered its way relentlessly north to Alexandria for its rendezvous with the Mediterranean Sea. One could see the felucca boats in the distance plying the river, carrying a constant stream of wide-eyed tourists who valued a peaceful, wind-powered journey down the Nile. Aside from the usual bustle of locals going about their business, Tahrir Square was quiet in those days. It was, and is, the gateway to the Egyptian Museum, which hosts spectacular exhibits of antiquity—mummies and other artifacts from the pyramids and various archaeological excavations around the country.

The oil and gas industry had brought great wealth to Egypt. Indeed, during the early days of our discussions with the Russians, we took delegations to Cairo to show them a resplendent example of how foreign investment could transform a country's economy. That said, a large segment of the population in Egypt still had not yet seen its standard of living improve dramatically.

Until 2011, when the contagion of the Arab Spring spread from Algeria to Egypt, the nation was tightly controlled by the iron-fisted regime of President Hosni Mubarak. The Egyptian leader displayed little tolerance for dissent and banned rival political parties such as the Muslim Brotherhood. However, under his leadership, Egypt had remained a staunch ally of the United States and Western Europe. His rule brought stability at a time when chaos prevailed in other parts of the Middle East, and the oil and gas industry had continued to develop and flourish during his tenure. That said, Mr. Mubarak's oppressive, and at times ruthless, style was increasingly at odds with the aspirations of the Egyptian people who wanted a greater say in their government and longed for the basic freedoms enjoyed by so much of the rest of the Western world.

While I was employed by Burlington Resources, several colleagues and I made a business development visit to Cairo in 2006. We met with two midlevel representatives of the US Embassy to be briefed on the extent of political risk, then and in the future, as well as other potential issues associated with operating in Egypt. Burlington was looking for potential locations to expand upon its relatively limited international portfolio. It had a 50 percent interest in a small undeveloped field in the eastern Nile Delta, operated by BP, but Burlington was interested in other opportunities in Egypt where *it* could operate. The embassy representatives related that the American government was "turning up the heat" on Mr. Mubarak to grant enhanced freedoms to his people. An avowed goal of the administration of President George W. Bush was to press for greater democratization in places like Egypt. When questioned, the embassy personnel conceded that the move could backfire if organizations like the Muslim Brotherhood took the opportunity to increase their presence in the Arab nation, or perhaps even seize control. It was a perceptive and prescient view of the turbulence that was to visit the nation five years later.

It is difficult for democracy to take root where the soils of freedom are shallow or nonexistent. Despotic regimes are more common than democracy in the Middle East, and lacking experience with Western democratic principles makes adopting them fraught with challenges. For most international business enterprises that are considering foreign investment in a host country, it is not so much the presence of democracy but rather a stable investment climate that is essential. Businesses rely upon clearly articulated and consistently administered laws and regulations.

We met with then–minister of petroleum Sameh Fameh and explained the important role that independents such as Burlington Resources could play in the exploration and development of Egypt's hydrocarbon resources. If the country were to rely on just a few large companies, those firms could exert considerable pressure on the government to achieve their aims. Competition, we explained, was a good thing. Big corporations had their role, but so did the medium-sized independents like Burlington. The minister and his colleagues seemed to agree.

Over the years, I had many dealings with Egyptian business colleagues. Most were friendly, engaging, enterprising, and smart. At this writing, the Egyptian people are still struggling to redefine and reshape their future, and the outcome of those struggles is still much in doubt. With the election of Abdel Fattah el-Sisi as president, they seem to have come full circle. While these are early days, his rule bears some striking similarities to that of Hosni Mubarak. It is my sincere hope that their collective aspirations can be met and a peaceful Egypt, with increased personal freedoms, will be the end result.

⊙ NEGOTIATIONS PRINCIPLES

A regime that understands and values the participation of foreign investors and that fosters a positive and predictable business environment is far more likely to attract investment from overseas. International companies can always choose to invest their capital elsewhere. In fact, since discretionary budget funds are not unlimited, most sophisticated multinational enterprises annually rank their potential investment projects so that they can decide which ones offer the greatest risk-weighted expected present value. Risk-weighting is crucial for any project, but particularly for foreign projects where a plethora of business risks, both technical and nontechnical, may come into play. The probability that a company will realize an attractive return on investment from Project 1 in a safe and predictable investing environment in Country A may be considerably greater than for Project 2 (which is identical in all respects except for higher risk) in Country B.

Where political risk is high and the government unstable, the chance that the foreign investor will actually obtain the benefit of the deal she negotiated may be quite low. An assessment of overall risk (not just political) is used by the company in determining what terms it will seek in its negotiations. Of course, one of the worst scenarios is for a previously

> stable country to become less so. The terms would have been negotiated during the period of stability and likely would not have envisioned the deterioration in the investment climate. The project could be at considerable risk from new taxes, unfavorable policies, or perhaps even expropriation. If the company has already made sizeable capital expenditures on the venture, it may be reluctant to withdraw. Furthermore, opportunities to sell for an attractive price may be severely limited.

Along for the Rio Ride

During 2007–2009, I worked for BG Group, the company that held the international assets of the former British Gas following the 1997 demerger. (The UK domestic natural gas business remained with the newly created entity Centrica.) I was based at Reading in the UK, BG Group's corporate headquarters, and held the title head of commercial.

One of my team's chief responsibilities was to lend commercial assistance and advice to the business units that had megaprojects in Australia, Kazakhstan, and Brazil. In South America, BG had an enviable position in four blocks in the Santos Basin subsalt play offshore from Brazil. The presence of a layer of salt can make "seeing" the hydrocarbons that may lurk beneath difficult, even with seismic data. The company publicly estimated mean total reserves and resources from subsalt Santos Basin to be about six billion barrels of oil equivalent, net to BG Group.[2]

In each of these blocks, BG partnered with Petrobras, the Brazilian state energy company. In its dual capacity as a quasi-governmental entity and as a partner, Petrobras played a somewhat murky and conflicted role. Because it had not been fully privatized, it was not clear that shareholder value maximization was its primary corporate objective. In addition, given its cozy relationship with the government, Petrobras wielded more power than its mere block ownership percentage would suggest.

Hence, BG was relegated to the unenviable position of trying to exert as much influence as it could over block operations in the Santos Basin. The company made heroic efforts to build solid business relationships with Petrobras. For instance, it set up joint working groups to consult on issues that might arise during exploration, development, and production. However, BG's influence was still quite limited and confined to where the interests of the parties were well aligned. I have referred to BG's relationship with Petrobras as akin to "being along for the ride." The plus was that they got to share in the benefits of block ownership, but the negative was that they were limited to sitting quietly in the backseat despite being well qualified to drive. Still, many other companies, which were left to eye BG's block position with envy from the sidelines, would gladly accept that role as opposed to not being in the vehicle at all.

BG's offices in Rio de Janeiro afforded great views of the city and the 30-meter (100-foot)-tall statue of Christ the Redeemer. Although prevalent street crime could mar one's visit to Rio, few cities offered so much rich culture. Visitors from the UK office were briefed upon arrival how best to avoid security incidents while in town.

⊙ NEGOTIATIONS PRINCIPLES

Brazil is another example of an "uneven playing field." As suggested above, Petrobras's influence in the decision-making process significantly outweighed its interest in a given block. To be clear, a firm's best leverage to secure key terms in a contract exists up front before financial commitments are made and spending has begun. However, where demand to enter an oil and gas basin is strong with plenty of competition, as was the case in the Santos Basin, even up-front leverage is severely limited. In the extreme, the competition bids away much of the economic rent, causing many companies to shy away from areas that are in heavy demand for precisely that reason.

Give and Take in the Caribbean

The nation of Trinidad and Tobago (T&T) lies in the southern Caribbean, a scant 20 kilometers (12 miles) off the coast of Venezuela and in close proximity to the Orinoco River delta. While Tobago is a popular diving and snorkeling destination, the main island of Trinidad has historically focused its attention more on agriculture and the energy industry than on tourism. Even though Trinidad possesses several idyllic beaches on its north coast, cruise ships tended to bypass the country for many years in favor of Barbados and the Lesser Antilles. Its hydrocarbon potential, however, has been a powerful attractant to companies like Amoco (and successor BP) for many years.

Amoco was instrumental in working with the T&T government to unlock its energy resources, most of which lay offshore to the east of the island. Oil was discovered and developed first, followed by reserves of natural gas sufficient in size to support construction of the Atlantic LNG (liquefied natural gas) facility now operated by BP.

In the fall of 1982, I was dispatched by Amoco to the island nation on an unaccompanied temporary foreign assignment. While the term may sound glamorous, it simply meant that I could not bring my wife along at company expense. I was to accomplish two goals: 1) to provide some additional training for the local economics team and 2) to support management in its ongoing negotiations with the T&T authorities. Amoco was in a "negotiations dance" of sorts—when the economic conditions were favorable, the government would be looking to increase its "take." Conversely, when conditions warranted (such as a low energy price environment), Amoco would approach the government seeking fiscal or other economic relief. As a consequence, it seemed to me that we were engaged in a perpetual negotiation. Both aspects of the assignment were intriguing, educational, and great fun, and I enjoyed the experience of living and working in the historic capital of Port-of-Spain where Amoco's offices were situated.

It took me some time to get accustomed to the heavy Caribbean accent with which the local employees spoke. I also harbored a fear that they might resent someone from the US coming to Trinidad to help increase their familiarity with project economics. My fears were unfounded, however, as I was, in fact, warmly welcomed. They wanted me to immerse myself in their culture, and I did.

On one occasion, when I was under the weather from a stomach ailment, Barbara, one of the more senior economists, said, "Come! We must go out to the Savannah and get you some coconut milk. That will settle your stomach!" The Queen's Park Savannah had been a famous landmark in Port-of-Spain since colonial times and was the site for cricket, equestrian events, football (soccer) matches, and other happenings. Small trucks were positioned around the periphery of the Savannah with loads of green coconuts heaped high in their rear cargo bays. Barbara seemed to know the vendor. He pulled out a machete and whacked the top off the coconut, revealing the channel to the liquid-filled inner chamber. He handed the opened fruit to me together with a straw. Smiling, he offered, "This should fix you up straightaway!" It did.

On the first day in the office, the chief accountant, an American, summoned me to his office and shut the door. He proceeded to tell me how dangerous Port-of-Spain was and implored me not to go out at night. While I am sure that his intention was to heighten my awareness of the surroundings, I felt that my stay in Trinidad would feel almost like a "prison sentence" if I were confined to that extent. I gradually introduced myself to various social events and nights out without incident. While I appreciated the chief accountant's admonitions, I am glad that I ventured out and partook of the local culture.

Still there *were* crime issues in Port-of-Spain, to be sure. My company apartment was located in the attractive suburb of Maraval. Construction of a new apartment building was going on nearby, and I could watch its progress from my living room picture window.

One Saturday afternoon, I was comfortably seated in the living room with my back to the window, fully immersed in the task of completing a jigsaw puzzle. Feeling a bit drowsy, I decided to head upstairs to my bedroom, which was located on the second floor of the duplex unit, for a nap. The windows in my apartment were fitted with bars to discourage break-ins. As I was about halfway up the stairway, I saw an arm protruding through the barred window in my bedroom. The would-be intruder was using a pole to lift my wallet off the bedside table and to maneuver it over to the window.

I yelled at the would-be thief. He dropped the pole and wallet, jumped down from the second-story window, and fled. I raced downstairs to see what had become of him, but he was nowhere to be found. Reflecting upon the incident later, I realized that I had acted instinctively and impulsively but perhaps not wisely. What if he had been armed? Although the thief escaped with nothing, I contacted the local police to file a report. I have no idea whether they ever captured a suspect, but I slept fitfully for the next few nights as I wondered if the intruder would attempt a return visit. I postulated that the break-in was perpetrated by a worker at the construction site since that location would have afforded him an unrestricted view. If I could watch them, I reasoned, they could see me, but this was just a working hypothesis. I never learned who had attempted the break-in.

It was now late November, and the island was preparing for Christmas and, more importantly for the Trinidadians, Carnival and the fetes that lead up to the main event held each year just ahead of Lent. Trinidad's Carnival is generally considered to be second only to Brazil's, and teams begin to assemble their costumes and floats months in advance of the grand parade. Unfortunately, my return to Houston was scheduled several weeks prior to Carnival, so I missed out on the glamorous and fun-filled event. However, I did attend a Christmas party at Amoco's Galleota Point facility on the southeast corner of the island. It was the operating base for the company's offshore production platforms. St. Nick, adorned in his customary red-and-white suit, arrived via helicopter from one of the offshore platforms, much to the delight of the boisterous assembled children. It was a rather bizarre scene—not the sleigh and reindeer that I recall from my upbringing at more northerly latitudes!

My weekly lesson plans for working with the local economists were nearly complete, and I had prepared a number of economic cases in support of the negotiations with the Trinidad government. The resident manager offered to let me extend my time in country. I thanked him but replied that it was time to return to my family. I extended my thanks to the local staff and management, made my rounds to say farewells, and headed back to Houston. The time had passed quickly, and I had enjoyed learning more about the T&T culture.

Seven years later in July 1990, an Islamist extremist group, Jamaat al Muslimeen, led by Yasin Abu Bakr, staged a coup attempt against the T&T government.[3] They stormed Parliament and took hostage Prime Minister A.N.R. Robinson and most of his cabinet. Robinson was bound and shot in the leg.[4] The insurgents also took over the Trinidad & Tobago television station, which was located near Amoco's offices in the TATIL Building.

The story circulating within Amoco at the time was that the insurgents also ordered a group of Amoco employees, who were in their offices, to line up outside against and facing a brick wall. The heavily armed extremists informed the employees that they would be executed and even went so far as to pull the triggers on some of the weapons, but fortunately the guns were not loaded. Thankfully, no employees were hurt, and the insurgents were eventually arrested (though they were subsequently granted amnesty and released). As of this writing, Abu Bakr remains a free man. The insurrection and attempted coup resulted in the deaths of twenty-four people and considerable destruction in downtown Port-of-Spain.[5]

I had long since left the Caribbean nation. The accounts related by the employees were nothing short of blood-curdling, however!

⊘ NEGOTIATIONS PRINCIPLES

"Government take" is a phrase used in the oil and gas industry, and undoubtedly many others, to signify the total benefit the government derives from a company's investment project. Components can include bonuses (sums paid upon award of a block or reaching other milestones such as production levels), royalties, taxes, and levies of various sorts. A company can compare and contrast how much of the generated cash flow from a hypothetical generic project it is able to retain in Country A versus Country B. This analysis allows it to determine the most favorable and unfavorable jurisdictions in which to conduct business.

I have found that governments frequently assume that a higher tax rate will equate to greater tax proceeds, but sometimes the reverse is, in fact, the case. As a regime raises the tax rate, it discourages investment. With fewer projects, total government take from aggregate foreign investment instead may fall. Even companies that have sizeable investments in a country and are unlikely to withdraw may slow or suspend future expenditures if their limited resources can earn a better rate of return elsewhere.

I have focused on situations where companies are in competition for award of exploration and development rights. Similarly, governments are in competition to attract foreign investment and must continue to provide adequate financial incentives for existing and incremental investments. The "negotiating dance" described above, in which the government of T&T and Amoco engaged, was designed to keep incentives in place at reasonable levels. The company needed to be able to demonstrate to its shareholders that it was earning an attractive and competitive return in T&T, and the government was accountable to its citizens to demonstrate that it was realizing its fair share of project cash flows from a depleting asset (the oil and gas reserves) which are then used to benefit the people. There needs to be transparency and accountability on both sides.

As this book went to press, there were allegations of corruption in Trinidad & Tobago. Afra Raymond, a leading expert on property valuation and project management, contended that he has been battling corruption in his native country of T&T. In a TED Talk, he pointed out some myths about corruption and noted that as long ago as 1982, two out of every three dollars earmarked for development was being wasted or stolen.[6] Little had been done since to address the need for financial transparency in governmental filings and accounts.

Trinidad and Tobago should be quite a wealthy nation following decades of oil and gas production, the "Kuwait of the Western Hemisphere," yet the average Trinidadian is not as well off as one would expect. Whether this is a matter of the government perpetrating corruption and fraud upon its own citizens, poor management of government take, or some other cause will be up to the appropriate Trinidadian authorities to assess. The country also needs to plan for the future with a more diversified economy as energy revenues inevitably tail off. At one point, the island nation relied upon revenues derived from oil and gas to meet 90 percent of its budget needs.

Danger in the Air

As an international negotiator, one practically lives in the air. A colleague once remarked that life on the ground was just "a long layover between flights!" My mother, who had studied physics, was concerned about my continual exposure to cosmic radiation. I can imagine that airline passengers are bombarded with more radioactive particles than would be the case on terra firma. However, air travel is the only practical way to conduct international negotiations, though virtual meetings via the Internet might avert the need for some travel these days. Face-to-face sessions, of course, are still the most effective means to establish and nurture professional relationships, so there will always be a need for them.

As we have seen from some of the earlier anecdotes, my line of work is not without its risks. I experienced the incident in Moscow with the youths who jumped me as I was on my way to McDonald's, the encounter with armed guards on the outskirts of Baku when AK-47s were aimed at me during a search and pat down, and the near-miss during the Litvinenko affair in London. One other memorable occasion bears mention.

In May 2001, while living in London and working for BP, I decided to rendezvous with my two teenage children in Denver for the long Memorial Day weekend in the US. Among other things, we planned to hike in the beautiful Rocky Mountain National Park north of Denver.

I was divorced at that point and I had to work out the details with my ex-wife, who was concerned about putting two unaccompanied minors on a plane bound for Denver while I was still en route to the Mile High City from London. I assured her that they would be fine, as the airline would be watching over them like a hawk.

The boarding of my American Airlines flight and takeoff from Heathrow Airport in London was routine and uneventful. As I frequently did, I gazed out the window as the English countryside, the Irish Sea, and then Ireland itself passed by below. About that time, the clouds finally obscured my view, and I lowered the shade in an effort to sleep (shut-eye almost always eluded me though on airplanes, even on long-haul flights). I planned to rest for a time and then watch a movie once the flight attendants began to serve lunch.

We continued to climb to our intermediate cruising altitude of some 30,000 feet. The ride was smooth, and the moving map showed that we had cleared the Irish west coast some time back and were now traversing the Atlantic south of Iceland. Meanwhile, in Houston, my kids, seventeen and fourteen at the time, would soon be preparing to head to Houston Bush Intercontinental Airport for their nonstop flight to Denver.

I gradually drifted off to sleep only to be awakened by an announcement from the captain. "Ladies and gentlemen, may I have your attention, please? We have a situation with smoke in the cockpit. Please do not be alarmed. We will be trying to return to Shannon Airport in Ireland. The flight there will take approximately one hour."

The word "*trying*" stuck in my craw, and I swallowed hard. Onboard fires were never welcomed, especially out over the icy-cold Atlantic with no place to land. The pilot continued, "We will be turning off all onboard electrical systems except those absolutely needed to fly the airplane. I apologize for any inconvenience." I doubted that anyone would be griping about a movie being interrupted.

The cabin fell deathly still while passengers contemplated their individual circumstances and potential fates. Only one lady in the row behind me cackled on in what seemed like an interminable monologue, though I guess her oration was aimed at the lady on her right. Being an analytical sort, I began to evaluate the situation. There were three ways I could die, I reasoned—from the impact, by drowning, or from hypothermia (or, I suppose, some combination of the three). I tried not to think about any of those options, though drowning seemed to me to be the least desirable.

My mind quickly shifted to my two children and how I wished I could hug them and tell them how much I loved them. I knew I had done so on many occasions before, but somehow it now struck me that I had not done it nearly enough. I longed to share so many things with them. I wanted to see them graduate from high school, get married, have kids, etc. I wiped a tear from my eye and stared aimlessly out the window. I could see no smoke, nor could I smell anything. I was seated in business class; this was one of the rare times that I was happy not to be in first class, closer to whatever might be burning.

Based upon the position of the sun, I concluded that the plane had reversed course already and was now headed east. I folded my hands, bowed my head, and prayed silently. I had not exchanged a word with the passengers seated in the middle or aisle seats, and I kept it that way, figuring that they wanted and deserved their own private contemplation time.

Then came another announcement from the cockpit. "This is the captain again," stating the obvious but sounding rather raspy. "My voice may sound a bit odd. That is because I am wearing a special suit and mask designed for these sorts of situations. We will be coming into Shannon at a low angle because we are not sure that our landing gear will deploy properly. If it does not, we will need to do a belly landing. The flight attendants will be around to give you instructions should we need to evacuate the aircraft upon landing." I thought, *Yes, landing is one of the certainties here, one way or another. But will it be a controlled soft landing, or will it be an unimaginable catastrophe?*

As the flight attendants parted the cloth partitions that separated the business-class from the first-class compartments and strode down the aisles toward me, I immediately read the terror on their faces. I then knew that the peril of which the captain spoke was very real and not exaggerated. One of the crew came over to a muscular man in the aisle seat in front of me. "Sir, you are seated near one of the exit rows and look to be pretty strong. When we land, I am going to need your help in pushing reluctant passengers down the slide. We cannot have a delay in exiting the aircraft, as the fire may be spreading by then."

I felt the aircraft descend gradually as if making a glider approach to the airfield. A glance at my watch confirmed that we were nearing the Irish coast (the moving map had been rendered inoperable, of course), but the dense clouds did not afford me a view. Sight of land would be most welcomed. Finally, we did clear the bottom of the cloud layer, which was quite low. I saw emergency vehicles in position, their lights flashing ominously. I had not heard the wheels descend, so I prepared myself for what I figured would be a *very* rough landing. My prayers were answered, though, and we touched down—a bit of a rough landing but without serious mishap. I recall the cabin filling with the sound of applause from relieved passengers. The American Airlines cockpit crew had handled the emergency very professionally.

Unfortunately, the scene on the ground was not handled as well. Although the ground staff was not expecting us, it seemed to me that they had had ample time to prepare for our arrival. My first thought was to contact my ex to advise her that our flight had been forced to divert to Ireland and would not be landing in Chicago as planned. However, I waited for what seemed like an inordinate period of time for the ground staff to announce what would happen to the passengers who found themselves unexpectedly at Shannon. Ultimately, we were advised that American's experts had examined the cockpit and nothing serious was found wrong with the aircraft (how could that be when smoke was reported on the flight deck?). The plane would be headed out shortly for the short hop over to Heathrow. We were given the choice of reboarding the original aircraft to London or remaining in Shannon and making our own onward travel arrangements. I made a quick call home and managed to preempt my kids' departure for Denver.

Trusting the airline, I reboarded the jet for what from all appearances turned out to be a perfectly normal flight. I was rebooked on a flight from Heathrow to Chicago where I was forced to overnight and catch an early connection to Denver the next morning. I rendezvoused with my kids, who had delayed their flight by a day, without further mishap. I hugged them both tightly and shared details of my misadventure at 30,000 feet over the Atlantic.

I knew that anyone could have boarded that flight; negotiators are not the only ones exposed to such risks. However, being a mathematician, I knew my odds were higher given the large number of flights I took year in and year out.

I enjoyed what was left of the weekend and prepared myself for my return to London and back into the fray. Is there risk in negotiations? Of course, but "no risk, no reward." Show me a line of work that is riskless, and I will show you one especially boring profession! International negotiations are anything but mundane, and I would not have it any other way.

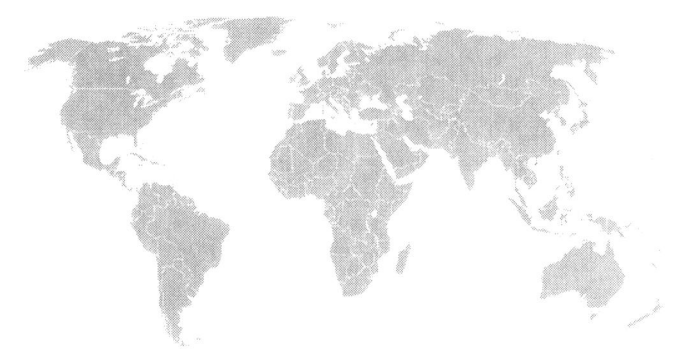

NINETEEN

The Only Constant in the Negotiating Environment Is Change

THE OIL AND GAS INDUSTRY, PARTICULARLY IN THE US, HAS long been regarded as a "boom or bust" business, prone to significant fluctuations in supply, demand, and price. Exogenous influences like energy cartels, wars, governmental energy policies, tax changes, and many other factors also play important roles.

Alternative sources of energy become more attractive as the price of oil increases but can be undercut again when it falls. It is not surprising that some exploration and production companies become complacent during the high crude price environment and borrow more than they should, only to be caught in a financial pinch when prices "head south." I am convinced that nobody is very good at forecasting energy prices over any significant period of time and certainly not at calling the inflection points where prices reverse course. While most companies test the robustness of their potential investment projects to a range of price and cost assumptions, picking a "base case" for prices is still a tricky proposition. Overly optimistic pricing assumptions at the time of project approval can mask the trouble that lies ahead when lower actual prices result in a significant drop in the realized rate of return.

Shale projects have greatly increased hydrocarbon production levels in the United States, allowing that nation to become the largest producer of natural gas in the world, beginning in 2010. According to the International Energy Agency and the Bank of America, the US also produced over 11 million barrels per day in the first five months of 2014, enabling it to surpass Saudi Arabia for the first time as the largest global oil and natural gas liquids producer.[1] Some oil and gas companies have redirected a larger proportion of their discretionary capital budgets to domestic projects, judging them to be less risky and perhaps more profitable than competing investments in foreign locations. This changing landscape affects the prioritization of projects and where and when negotiators are deployed.

Much of the success with shale production has been attributed to horizontal drilling advances and hydraulic fracturing (or "fracking"), a technique used to stimulate production. Drilling fluids are forced under high pressure into reservoir rocks, fracturing the formations and allowing oil and gas to flow at greater rates into the well bore.

Of course, the "shale revolution" is not without controversy. There are spirited debates in the US as to whether or not to permit fracking. Communities and governments want to ensure that groundwater will not be contaminated with the chemicals used in the fracking process.

There is some opposition in the US as well to energy exports, and governmental energy policy remains a polarizing topic. Then, too, there is price uncertainty. As the Chinese economy began to cool and Europe was suffering through a difficult economic cycle, the world crude price began a swoon in the second half of 2014 and into 2015.

Shale drilling is a relatively expensive proposition, and the production from wells can decline quite rapidly, requiring continuous drilling to maintain the production plateau for a field. Amid growing signs that the global supply of oil was outstripping demand, OPEC (under the leadership of Saudi Arabia) in 2014 chose not to reduce its production, aggravating the oversupply situation, thereby accelerating the oil price decline. Perhaps OPEC was concerned about the growing US share of global production. Since Saudi production costs (estimated by many to be about US $10 per barrel) are well below those of the shale producers, the kingdom can continue to produce at prices that would be unacceptable to North American drillers. Accordingly, drilling rigs increasingly have been idled in the US where the rig count dropped 46 percent between October 2014 and March 2015.[2] Indeed, the US claim to top global oil producer may be short-lived.

Further complicating the investment picture in 2014–2015 was increasing global unrest, not just in the Middle East, where it seems to be endemic, but in North Africa, Nigeria, and Ukraine. Relations between Russia and the West are at a post–Cold War low. Might Cold War II be just around the corner?

As I reflect on my negotiations in the former Soviet Union, I am struck by how Russia and Azerbaijan have chosen to take such divergent paths. During the 1990s, Russians rejoiced in their newfound freedoms of speech, press, and assembly. One remarked to me during a break from a negotiating session, "We could never joke about the government in the past. Now we can." In present-day Russia, however, those who have chosen to speak out in opposition to the government have quickly come to appreciate that criticism of the Kremlin is neither valued nor tolerated.

Many international oil companies have been forced to leave Russia. Flush with petro-dollars, the Kremlin felt that it no longer needed foreign investment or advice. Most of the energy sector has now been consolidated under governmental control, reversing the decisions of the 1990s when semiautonomous companies were created out of the vestiges of the oil and gas production associations.

The story in Azerbaijan could not be more different. The 1,768-kilometer (1,100-mile) BTC Pipeline became fully operational in 2006.[3] At an estimated cost of US $3.7 billion, the pipeline began transporting crude oil from Azerbaijan to the Mediterranean port of Ceyhan, Turkey, from which it was then transshipped to world markets. After initial reservations, the authorities in Azerbaijan chose to trust the international companies who promised that their investments in the country would completely transform the former Soviet Republic's economy and raise the standard of living of its citizens. Those promises have become reality.

From time to time, I also think back to the discussions my colleagues and I had in Georgia with President Eduard Shevardnadze in the early 1990s about the notion of an export pipeline traversing his nation. The Georgian head of state died on July 7, 2014, well after completion of the BTC pipeline.[4] He must have been particularly proud of his country's role in that megaproject; Georgia continues to benefit financially from pipeline transit fees.

In the end, probably the most rewarding aspect of negotiations was realizing that I had contributed in some small way to enhancing the quality of life in places like Azerbaijan. It all started with a presentation in 1991 to a USSR-wide television audience and the negotiations "across the table" that followed. Although Azerbaijan may be one of the most striking examples of the benefits of collaboration between multinational companies and host governments, others abound, such as Egypt and Papua New Guinea. The Ok Tedi mine provided training and employment to local citizens in PNG's Western Province, allowing them to lead lives they never dreamed possible.

APPENDIX

Negotiations Principles
A Reference Guide

THIS APPENDIX IS INTENDED TO SERVE AS AN EASY REFER-ence to the negotiations principles covered in the book (and it adds a few that were not). The principles, provided in outline format, are grouped under the following headings:

1. Some Overarching Principles / 209

2. The Macro-Environment—Backdrop to the Negotiations / 210

3. The Effect of Competition / 210

4. Know Your Opposition / 211

5. Building Trust with the Opposition / 212

6. Relative Power of Negotiations Participants / 212

7. Composition of Your Negotiating Team / 214

8. Be Aware of Your Own Metes and Bounds / 214

9. Preparing for and Introducing the Negotiations / 214

10. Establishing the Negotiations Playing Field / 216

11. Techniques at the Table / 216

12. Economics of a Deal / 217

13. Balanced Deals and Deal Stability / 218

14. Driving to Close the Deal / 219

15. Potential Pitfalls / 220

16. Qualities of the Successful Negotiator / 224

17. Negotiating Styles / 225

18. Security of Information / 226

19. Joint Ventures / 226

20. Digesting, Analyzing, and Capturing the Learning Experience / 227

1. **Some Overarching Principles**
   a. Only fair and equitable deals endure and pave the way for repeatable business transactions
   b. Goal should be to obtain the optimal risk-weighted present value of the deal for you or your company, as appropriate
   c. Always be willing to walk away from a deal
      i. If you signal that you must conclude the deal no matter what the cost, the cost will be more than it could be
      ii. If you must do the deal, at least give the appearance of a willingness to walk away
      iii. Inappropriate or inconsistent body language or a slip of the tongue can erase an otherwise convincing and sincere acting job
      iv. Be prepared to walk away at the bazaar
         1. If the seller chases you, there exists a final chance to conclude the deal at a more desirable price
         2. If he does not, the asking price was too low or he is insufficiently motivated
      v. Frequently a negotiator working for a corporate enterprise would be willing to walk away, but his company would not, and this point is well understood by the counterparty to the negotiations. This misalignment needs to be corrected through an open and frank discussion with management. It is rare that any project is so important that a company cannot justify walking away if the minimal terms cannot be achieved.
   d. Sometimes the best deal is no deal (the asking price is just too much or the assumptions made prior to the onset of negotiations are way off the mark, resulting in a set of real parameters that are unacceptable)
   e. Develop mutual trust, the underpinning to any successful business relationship, with your counterparty

f. Adhere to all applicable laws, maintain high ethics and transparency, and avoid any appearance of conflicts of interest or other impropriety

2. **The Macro-Environment—Backdrop to the Negotiations**
   a. Negotiations do not play out in a political vacuum
   b. If it is an international deal, what are the geopolitical considerations (e.g., the ability of a country to export product to world markets without provoking another country that might intervene to cut off those flows)?
   c. What relationship does your country have with the host country? For example, are there bilateral tax treaties?
   d. Does the host country have a long tradition of honoring agreements and allowing equitable mechanisms to resolve contractual disputes?
   e. Countries that respect the rule of law and provide for a predictable investment climate are far more likely to attract foreign investment. A stable investment environment is usually more important to companies than the particular form of government. Among other things, governmental risk can range from new taxes and levies, to increased state interference in decision making, to creeping or outright expropriation
   f. Does the host country allow arbitration in neutral jurisdictions and unrestricted repatriation of capital?
   g. Are there any restrictions on the ability to repatriate cash flow?
   h. If the deal is a domestic one, are there opposition groups that must be engaged and won over to support (or at least not block) the project?

3. **The Effect of Competition**
   a. Competition helps ensure that a good or service is priced efficiently and appropriately in the marketplace

b. If there is too much demand for an item, the ability to negotiate may be limited (excessive demand and limited supply forces up the price)
c. Excessive competition in a bidding environment can translate into such an erosion of potential profit that the "winner" winds up with a project that never gets developed since it has insufficient remaining economic value (sometimes companies take this approach merely to get their foot in the door, hoping to renegotiate the terms at that point, but this unethical technique can lead to an evaporation of any trust existing between the parties)
d. Where the government controls the price or other terms, flexibility in negotiations may be severely limited
e. Sometimes differentiating your product or service from others on offer can allow you to separate yourself from the competition and perhaps obtain a price premium

4. **Know Your Opposition**
   a. What are their drivers?
   b. How motivated are they to do the deal, or are they personally "at risk" if they finalize a transaction that might be criticized by their organization?
   c. Try to place yourself in their shoes to understand the importance of concluding the deal (this step may help you to anticipate their moves)
   d. Understand which elements matter most to them and which they would most highly value
   e. Who are the key decision-makers in their organization?
   f. How much authority does the counterparty have (must they go to their boss for approval, or are they empowered to make the key decisions)?
   g. Does the counterparty have the background, experience, and expertise to understand the deal that is being contemplated? If not, do they need competent advisors to represent them in the discussions

so that both sides are speaking a common business language? To avoid the appearance of a conflict of interest, funding for such advisors/legal counsel should not come from your organization

5. **Building Trust with the Opposition**
    a. Trust takes a great deal of time and patience to build, but it can be squandered in an instant
    b. Build professional relationships and establish trust well *before* they might be tested. Doing so builds a repository of goodwill that can be drawn upon during a contentious negotiation or when trying to resolve a dispute
    c. Do not confuse relationship-building with hard-nosed negotiations. It is possible to engage in a difficult negotiation while maintaining, or even improving, a business relationship
    d. Similarly, building trust should not be confused with a conciliatory approach to negotiations (many cultures respect a tough but fair negotiator, not someone who arrives on Day One waving the white flag)
    e. No substitute exists for some face-to-face meetings when possible (can be supplemented in the interim by video links, phone calls, and exchange of electronic communications and documents)
    f. Understand and respect differences in culture, but do not sacrifice your values in trying to honor them
    g. Use of threats (overt or implied), failure to keep your word, vacillation in positions, and use of a "bait and switch" approach to negotiations all can erode or destroy trust
    h. Use of external benchmarks or standards can help remove suspected bias in deals

6. **Relative Power of Negotiations Participants**
    a. If parties are not on a level playing field, it can profoundly affect the outcome of a negotiation (see the Egyptian camel story—once you are up on the camel and then trying to negotiate the fare [ransom?]

to be lowered back to the ground, your negotiating position is substantially weaker than before you were hoisted aboard)
b. Negotiating leverage in an investment project is usually greatest up front before key economic terms are agreed upon, an agreement has been signed, or any financial commitments have been made (the willingness to commit financial resources is a significant lever that many negotiators seem to underappreciate)
c. Never proceed with a project before a comprehensive deal has been negotiated, signed, and enacted into law (if appropriate), no matter how appealing the short-term benefits associated with an interim arrangement might appear—the likely outcome is a greatly diminished negotiating position for concluding the longer-term comprehensive project agreement
d. Where demand for a good or service significantly outstrips supply, the negotiator who is trying to strike a deal at less than full price will likely be disappointed
e. A negotiator can sometimes unnecessarily place himself in a position of weakness by "negotiating with himself/herself" (giving in on terms in anticipation of the counterparty's negative response) or by adopting a conciliatory stance in negotiations. Await the counterparty's response, the contents of which may sometimes come as a pleasant surprise
f. The aggressive negotiator frequently bests his conciliatory opponent (though continued use of this approach can result in stiffening of the resolve of the other side and a slowing of the negotiations process as positions become entrenched)
g. Time pressure (real or imagined) on one party to conclude a deal, especially when spotted by the counterparty, can lead to a weakened negotiating position and a suboptimal outcome
h. If a negotiator is in a relatively powerful position, does the counterparty understand that? If he/she perceives the power disparity differently, it may not be easy to exploit the advantage

7. **Composition of Your Negotiating Team**
    a. Do you have the necessary expertise at the table (or is it easily accessible)?
    b. Be sure to seek input of law, tax, economics, and accounting expertise early on when a deal is first shaped since the flexibility to optimize a deal later may already be compromised
    c. Always good to have more than one set of eyes and ears at the table
        i. Others can act as witnesses
        ii. They can spot and interpret subtle signals coming from across the table
        iii. They are in a better position to judge how effectively the discussions are going since the negotiator is frequently too engaged in the actual conversations
    d. Have a pragmatic lawyer at the table with you

8. **Be Aware of Your Own Metes and Bounds**
    a. What are the essential elements of the deal from your organization's perspective, and what is the real value proposition?
    b. Where can you show flexibility without compromising the core economics of the deal?
    c. Will the deal set new or adverse precedents?
    d. How does the proposed arrangement square with existing company policies?
    e. How much flexibility do you have to restructure the terms or reshape the deal?
    f. How much support is there for the deal within your own organization?

9. **Preparing for and Introducing the Negotiations**
    a. Develop a plan in advance of how you hope the negotiations will go (may be more like a decision tree—there may be several possible

outcomes when an event occurs; the plan would contain a proposed path to take for each eventuality). Plans need to be flexible and be revised as you go (things rarely go as you would foresee them, but a flexible plan is far better than none at all)
b. Do your homework—gather as much intelligence as you can about your counterparty and his/her organization, their motivations, strengths and weaknesses, and what deals they have struck in the past
c. Develop a good understanding of the value of the project or asset that is the subject of the negotiations, including hidden value (for example, a premium for control or operation of a project)
d. Know your bottom line (the most one is willing to pay if he/she is the buyer, or the least one is willing to accept if he/she is the seller)
e. Calculate the value of potential concessions/gains
f. Have a list (not visible to the other side!) of your break points, show-stoppers, or must-haves
g. Understand your next best alternative to doing the proposed deal (sometimes it is an alternative investment or perhaps even doing nothing; rarely is there no realistic alternative), and assess what the counterparty's alternative would be
h. How are you going to introduce and frame the discussion? This step needs to be carefully thought through and should be developed in a manner calculated to be most convincing to the other side (or at least not easily refutable)
i. Rehearse your introduction and negotiations plan in advance for colleagues to critique
   i. Can both sides agree on a common goal, even if a fairly broad one? Good to have a proposed agenda for meetings and agree how much time is available for the meetings

10. **Establishing the Negotiations Playing Field**
    a. Positioning the "goalposts"
        i. Thoughtful initial offer
        ii. Appropriate counteroffer
    b. Recognize that the starting offer made by a seller may bear little relationship to the intrinsic value of the good or service on offer (the seller may hope that you would agree to split the difference, potentially leaving her with a sizeable profit margin if you have not probed to get a better sense of the asset's true worth). If an item is worth 20 euros, the initial offer from the seller is 100 euros, and you split the difference at 50 euros, you have paid her 2½ times what the item is worth, leaving considerable money on the table
    c. Never accept the first offer made by a counterparty because it is unlikely to be anywhere near her bottom line
    d. Counterproposals are most effective if they have logic, reason, and objectivity on their side (remember that fair deals are the ones that endure)

11. **Techniques at the Table**
    a. Usually keep emotionalism out of the process unless intended for conscious effect—in some cultures, emotions are regarded as weakness or as unprofessional behavior; in other places it is interpreted as sincerity and compassion to reach a successful outcome
    b. The flinch—show shock if the terms disclosed are not what you expected (perhaps flinch even if they *are* in order to negotiate even better terms)
    c. Feigned outrage
    d. Use of implied threats (should be used sparingly; better to keep the discussions on a positive footing)
    e. Controlled messaging
    f. Active listening—do not talk to the exclusion of listening

g. Breaks—call "time out" if you need to confer with team members to assure you are all aligned (huddle in a separate location where you cannot be overheard)
h. Identify items that may be disproportionately valued by the other side that can be fashioned into a win/win outcome
i. Look for ways to expand the "pie" of economic rent available from a project
j. Someone other than the negotiator should keep accurate and detailed notes
k. Whenever possible, control the documentation process (your side prepares the initial and all subsequent drafts of the agreements). This allows your side a tactical advantage so long as the drafting is done reasonably and the advantage is not abused
l. In most cultures, eye contact (if not excessive) can send a message of respect, sincerity, and fair dealing

12. Economics of a Deal
a. Generate a present value table that shows the worth of each component of the deal so that you know what you are gaining or conceding during the negotiations
b. Can be useful to get a third party's estimate of the value of an asset you are buying or selling to ensure that you are not overlooking a source of value
c. Remember that it is beneficial to defer costs and accelerate revenues as a result of 1) the time value of money (a dollar received today can be reinvested to be worth more than a dollar ten years down the road), and 2) the longer the period of time it takes for a project to reach its targeted rate of return, the greater the likelihood that some unforeseen event will intervene to frustrate your best-laid plans
d. If the negotiation concerns an enhancement or expansion to an existing project, examine the incremental cash flow (over and

above what you would have received anyway) and economic parameters to understand the true value associated with the additional investment

e. Always look at net cash flows after all taxes. Taxes are just another cost of doing business

f. Sunk costs should be disregarded when assessing point-forward economics unless they have some financial impact on future cash flows (e.g., prior expenditures may have generated tax losses that might be carried forward and applied to reduce future taxable income)

g. Check to make sure all opportunities to exploit disproportionate valuations have been utilized

h. Examine risk-weighted economics since projects usually are exposed to myriad different risks

i. Always be on the lookout for options that you can incorporate into a deal; options always have a nonnegative value. Sometimes options that are seemingly trivial or considerably "out of the money" wind up being of major importance as conditions change over the life of a project

j. Recognize that some state entities are more driven by governmental strictures than by a profit motive. A state entity searching for hydrocarbons around the globe may be more interested, for instance, in accessing a secure supply than in the price they pay or the economics of the transaction

k. Since most corporate budgets are limited, it is useful to rank projects to select those that offer the brightest economics prospects. Do not rely on internal rates of return for this exercise, as they can result in incorrect conclusions

13. **Balanced Deals and Deal Stability**
    a. Deals that place one side or the other at a significant economic disadvantage rarely last. Seek a deal that the counterparty would

reasonably conclude is in its economic self-interest and that it could "sell" as a good outcome back home. Would you be willing to accept the deal that you are asking the opposition to accept?
b. Unfair deals undermine trust and threaten the ability to conclude future business transactions
c. Involvement of multilateral entities like the World Bank/IFC or EBRD can help provide stability to a deal and ensure fairness, but they can also slow the process down
d. It is important, wherever possible, to have the final agreement enacted into law by a country's legislative body
e. Economic equilibrium clauses can help in the event that the fundamental fiscal assumptions change after the project is underway
f. Normally, the project sponsors must take a reasonable degree of risk (such as making up-front investments at their sole risk) to be allowed to earn a corresponding rate of return. The most a party should expect is the chance to earn a fair rate of return without the rules of the game being altered following the up-front negotiations—i.e., the opportunity to benefit from the deal that was struck without adverse governmental intervention
g. Make sure ambiguous provisions have been eliminated from any agreement and that all relevant disciplines have had ample time to review the provisions that fall into their areas of expertise. Ambiguous provisions invite disputes and litigation. Take the time necessary to thoroughly vet the document, asking questions of the counterparty where necessary

14. **Driving to Close the Deal**
    a. Seek opportunities for win/win (more likely to result in repeat business than win/lose, which is a zero sum game—my win is your loss)
    b. Enlarge the "economic rent pie" whenever the opportunity exists, and recognize items which may be disproportionately valued by the other side

c. Close the gap (remembering that the midpoint may be the wrong place to wind up); creativity may be required, especially if the gap is large
d. It is sometimes advantageous to group unresolved issues into a single proposal to resolve (two parties might quibble over the individual values of items a, b, and c, but when the components are grouped together, there is at least a chance that an agreement can be reached on the overall package)
e. Differences in the parties' assessment of risk (or risk tolerance) or the time value of money may present an opportunity to fashion a solution to closing the gap

15. **Potential Pitfalls**
    a. Individual/Team Issues and Behavior
        i. Avoid unintentional signals (verbally or via body language) at the negotiating table
        ii. Refrain from taking inconsistent (or frequently changing) positions or backtracking on previously agreed principles or terms
        iii. Never respond to "What are you prepared to pay?" The real question is, "What is an item worth?"
        iv. Avoid claiming a proposal is your final offer unless it really is and you are prepared to walk if you do not get it
        v. Be aware of the opposition trying to "forum shop" (if they do not get the answer they want from you, they go over your head or around you)
        vi. Do not agree to terms on a one-off basis; consider individual terms to be "for discussion purposes only" until all the key terms are agreed as a package; this approach avoids the other side "cherry picking"
        vii. When selling an asset, never say, "I could accept a price for my donkey of between $15 and $25." Translation: You have just

offered to sell it at the low end of the range. Offers and counteroffers should be precise unless there is a real reason to create ambiguity. In most cases, ambiguity sits squarely on the quick path to misunderstandings

viii. Beware of the "unempowered" car salesperson tactic (they waste your time negotiating, only to divulge later that any deal needs to be run by the boss who is the real decision-maker and who will try to extract additional concessions from you). This can best be remedied by confirming early on that you are dealing with the empowered representative. This tactic should not be confused with the caveat that many deals require final approval of the company's board of directors, which is usually legitimate if the deal is large enough, in a new country, or otherwise requires such approval

ix. It almost goes without saying, but never negotiate while mentally impaired (including from a lack of sleep)

x. Avoid negotiating in a language other than your native one unless you are fully fluent in the other language. If you are negotiating in Moscow and are somewhat conversant in Russian, it is still best to insist on a translation into your native language. It gives you added time to think before responding, and the other side, believing that you are not a Russian speaker, may let down its guard and mention something in its language that could be useful to you in the negotiations process. I am not suggesting that you be untruthful when asked if you speak the language, but consider not volunteering that fact

xi. Try to get the official language of the agreement to be your own

xii. Viewing a response from the foreign counterparty exclusively through the lens of your own culture may be a mistake; in some cultures, "no" does not really mean no, and "yes" is not an unqualified yes

xiii. Do not assume because a counterparty says the terms of a deal are "nonnegotiable," that there is really no flexibility. Everything is almost always negotiable to varying degrees, at least around the edges
xiv. Do not rely upon oral discussions not backed up by a written record (label documents that are circulated "for discussion purposes only" unless agreement has been reached on all of the key terms). Even then, it is appropriate to state that everything is subject to concluding a mutually acceptable comprehensive agreement
xv. Core economic terms can be listed on a term sheet and initialed. However, in many cases, a Heads of Agreement or Letter of Intent is an unnecessary step. Such documents, by their very nature, are broad statements of principle, incomplete and frequently unenforceable. If there has been sufficient discussion and agreement in principle around the key terms (economic and otherwise), it is usually desirable to jump directly to negotiating a draft agreement

b. Organizational Issues and Behavior
i. Lack of continuity—Organizations should ensure an effective hand-off of a negotiation if the lead negotiator moves on
ii. Avoid sacrificing accountability by transferring negotiators prematurely (accountability should transfer with them; otherwise, a negotiator may be less motivated to ensure that any deal that is reached is both workable and in the company's long-term interest)
iii. Do not undercut your negotiator—Top management should support its lead negotiator and only step in to resolve a few outstanding issues at the end of a negotiation (to enter the fray prematurely only serves to undermine the authority of the lead negotiator and commit the senior executive to ongoing

involvement that he/she probably would not welcome)
   iv. Avoid placing arbitrary and unrealistic time deadlines on negotiators since they usually work to the advantage of the counterparty. Sometimes establishing a reasonable deadline that applies to both parties can serve to propel a deal forward to a successful conclusion, but it is rare that both parties feel the same degree of urgency. Governments tend to be far more patient than private sector entities; governments are content to wait out their private sector counterparties, figuring that they will get a better deal (and they frequently do)
c. Other Issues
   i. Use care to avoid setting precedents in negotiating on one topic that may carry over and hurt you on another. For example, if a decision is reached to use the host country's law and courts (avoid when you can!) on some relatively inconsequential agreement, the counterparty may contend that you have implicitly signaled a willingness to follow the same course of action in a more comprehensive and far-reaching agreement
   ii. Identify and avoid situations where conflicts of interest may arise. In some countries, a state entity may be your coventurer, your supplier of contract services, the purchaser of your production, the state regulator, etc. The value derived from some of these ancillary roles may dwarf in value the cash flow derived from its role as coventurer. This situation diverts the focus away from maximizing the value of the project
   iii. Avoid introducing unintended incentives that encourage parties to take positions that are adverse to your project. A "cost plus" contractual structure does not incentivize the contractor to control costs; quite the opposite, actually
   iv. Be wary of potential partners or counterparties who are overly litigious. There are unscrupulous parties for whom lawsuits

are a greater source of income than the projects in which they are involved! Having to defend against legal claims can also be highly disruptive to a business and certainly can quickly destroy a working relationship

16. Qualities of the Successful Negotiator
   a. Not all personalities are well suited to negotiations, and corporations frequently thrust an employee into the role for the first time with little or no training (many negotiators learn from being thrown into the swimming pool at the deep end while onlookers implore them to swim!)
   b. At a minimum, the negotiator must demonstrate sound ethics and a willingness to strike a fair deal. (Ensure that you adhere to all applicable laws and regulations, including the terms of legislation like the US Foreign Corrupt Practices Act and the UK Bribery Act, and hold your partners and counterparties to the same standards.)
   c. Strive to be an articulate leader who can effectively describe how the project could benefit both parties
   d. Do not share too much information at the table. The most obvious examples are i) not sending the signal that a deal must be done at any cost and ii) not unwittingly disclosing to the other side one's bottom-line authority (e.g., the maximum amount a company authorizes its negotiator to pay at an auction)
   e. Be a good listener who can empathize with the counterparty's position (much can be gleaned from "active listening" and sometimes letting the counterparty present its position first)
   f. A successful negotiator appreciates that sometimes the path of least resistance to agreement (and therefore the quickest path) may not be the straight-line route
   g. Be someone who has a good sense of the "big picture" and never loses sight of the value proposition that makes the deal attractive

h. The negotiator should avoid becoming emotionally involved in doing the deal (best to have someone negotiate who is not the project developer/sponsor or reports directly to them). The negotiator should be looking out for the interests of the corporation and its shareholders, avoiding a more provincial or narrow view. Having a performance appraisal prepared by the project sponsor can undermine the independence of the negotiator
   i. A good negotiator recognizes that negotiations are a team sport and that he/she benefits from the timely inputs from experts. Be sure to thank your team for their valuable inputs and give constructive feedback where appropriate

17. **Negotiating Styles**
    a. Hard-nosed/uncompromising— May be appropriate if the counterparty has no realistic alternative to the proposed transaction
       i. Most effective when matched against a conciliatory/nonaggressive counterparty
       ii. Does not work well if the opponent does not recognize the power advantage of the negotiator's side or if the opponent is a hard-nosed, uncompromising negotiator himself
    b. Conciliatory/cooperative— May work well if this style is matched across the table
       i. May allow a quick agreement if the opponent is hard-nosed/uncompromising, but outcome may be suboptimal
       ii. If both parties are conciliatory, there could be an absence of leadership at the table
    c. Reliance on external benchmarks and standards
       i. May be easier to defend positions since they are based upon reasoned positions
       ii. Requires agreement on both sides as to what are appropriate benchmarks and standards

d. Negotiators can and should modify their approach to fit the circumstances and the personality across the table (really tests the skill of the negotiator who probably has a comfort zone in a given style)
   e. Styles also differ by level of inclusivity—does the negotiator believe it is a team sport (as I do) or a one-man/one-woman show? If it is the latter, it is critical that the negotiator seek buy-in and approval of experts behind the scene, especially in a larger organization

18. **Security of Information**
    a. "One peek is worth two finesses," as they say in the game of contract bridge. Avoid leaving work papers, notes, data files, and the like where they can be compromised or intercepted
    b. Lock away your papers before any recess (like breaking for lunch), leaving a trusted guardian near the entrance to the room when practical
    c. Conduct the most sensitive oral discussions during walks outdoors away from buildings
    d. Ensure that the loyalty of interpreters/translators is beyond doubt (aside from the security aspect, these vital communication links must understand the subject matter sufficiently so that translations are not too literal, losing the real meaning in the process)
    e. Meet in neutral locations if the danger of electronic eavesdropping is present

19. **Joint Ventures**
    a. Joint ventures (JVs) are dependent upon an ongoing alignment of interests; even if they are aligned at the outset, they can easily diverge with time
    b. If a corporate entity is formed, parties should refrain from selling products or services to the entity unless it can be clearly

demonstrated that the transaction prices are true arms-length ones that are on par with other comparable alternatives in the market
c. The rights and obligations of the parties need to be clearly set out in a shareholders agreement or joint operating agreement, and the voting procedure and mechanism clearly understood and agreed by all
d. Cultural differences can complicate JV relationships (e.g., in some cultures a party may need to save face, which could reduce the universe of options available to reach agreement with the counterparty)
e. If a negotiator is acting on behalf of the partnership/consortium/corporate entity, his/her positions should be agreed separately ahead of time and the negotiator's remit made clear
f. In a competitive tender environment, a bidding agreement should be agreed among the parties at the outset, which allows one or more parties to drop out under certain circumstances and others to proceed. The bidding agreement would be replaced, in the event of a successful bid, with a shareholders agreement or joint operating agreement
g. Sometimes a minority participant in an agreement carries more weight than its stakeholding would suggest. A small interest held by an entity with close ties to the government can frustrate the will of the majority in a consortium. Where possible, it is useful to avoid partnering with such entities

20. Digesting, Analyzing, and Capturing the Learning Experience
a. Every negotiation offers opportunities to learn and improve your capabilities as a negotiator—what worked well, what did not, and where is there room for improvement?
b. Did you use the right style of negotiations for the circumstance? Did you understand and fulfill the counterparty's needs and aspirations in the deal?

c. The team should keep good notes as you go through the process, and you should prepare a retrospective report when the transaction has been completed (or talks have broken down). These notes not only will be useful to you in the future but also can assist whoever picks up your project/negotiation after you move on
   d. If the discussions were discontinued, what were the primary reasons? If they succeeded, is the deal a balanced one where both (all) parties can feel good about the outcome?

Notes

Introduction

1. Dan Morgan and David B. Ottaway, "Azerbaijan's Riches Alter the Chessboard," *Washington Post,* published October 4, 1998, accessed April 14, 2015, A1, http://www.washingtonpost.com/wp-srv/inatl/europe/caspian100498.htm.

Chapter 1

1. "History—The Flight of Apollo—Soyuz," *NASA,* accessed February 4, 2014, http://history.nasa.gov/apollo/apsoyhist.html.
2. "World War Two Casualty Statistics," *Second World War History,* accessed February 4, 2014, http://www.secondworldwarhistory.com/world-war-2-statistics.asp.
3. John Lavitt, "New Study Reveals Drinking's Dark Toll On Life Expectancy In Russia," *The Fix,* published February 20, 2014, http://www.thefix.com/content/new-study-reveals-drinking%E2%80%99s-dark-toll-life-expectancy-russia.
4. "Baku (Baki): History," *Lonely Planet,* accessed April 24, 2014, http://www.lonelyplanet.com/azerbaijan/baku-baki/history.
5. Mir Yusif Mir-Babayev, "Azerbaijan's Oil History: A Chronology Leading up to the Soviet Era," *Azerbaijan International Magazine* (Summer 2002), accessed April 24, 2014, 34–40, http://www.azer.com/aiweb/categories/magazine/ai102_folder/102_articles/102_oil_chronology.html.

Chapter 2

1. Urja Dav, "Edwin Drake and the Oil Well Drill Pipe," *The Pennsylvania Center for the Book—Edwin Drake and the Drill Pipe,* published Summer 2008, http://pabook.libraries.psu.edu/palitmap/DrakeOilWell.html.
2. A. A. Narimanov and Ibrahim Palaz, "Oil History, Potential Converge in Azerbaijan," *Oil & Gas Journal,* published May 22, 1995, http://www.akifnarimanov.com/akif/extdocs/OGJ_19950522.pdf.
3. *The World is Not Enough,* Directed by Michael Apted (1999; United States: MGM Home Entertainment, May 16, 2000), videocassette.
4. Narimanov and Palaz, "Oil History, Potential Converge in Azerbaijan."
5. Dom Joly, "Amazing Azerbaijan: Baku to the Future in the Capital City at the Very Edge of Europe," *The Daily Mail,* last modified July 28, 2010, http://www.dailymail.co.uk/travel/article-1297564/Amazing-Azerbaijan-Baku-future-capital-city-edge-Europe.html.

Chapter 3

1. Stephen Kinzer, "The Fallen Commissars of 1918, Now Fallen Idols," *The New York Times,* published September 9, 1997, http://www.nytimes.com/1997/09/09/world/the-fallen-commissars-of-1918-now-fallen-idols.html.

2. Bob Dole, *One Soldier's Story: A Memoir* (New York: HarperCollins, 2005), 234–37.
3. Jonathan Yardley, "Bob Dole's 60-Year War," *The Washington Post*, published April 12, 2005, http://www.washingtonpost.com/wp-dyn/articles/A45452-2005Apr11.html.
4. Dole, *One Soldier's Story: A Memoir*, 234–37.

Chapter 4

1. "History of Azerbaijan," *Azerbaijan America Alliance*, accessed April 1, 2015, http://azerbaijanamericaalliance.org/history.
2. "History of SOCAR," *Republic of Azerbaijan*, accessed June 9, 2014, http://www.azerbaijan.az/_StatePower/_CommitteeConcern/_committeeConcern_e.html.
3. *Pennzoil Exploration and Production Company v. Ramco Energy Limited*, United States Court of Appeals, 5th Circuit, May 13, 1998, https://law.resource.org/pub/us/case/reporter/F3/139/139.F3d.1061.96-20497.html.

Chapter 5

1. Joseph A. Kechichian, *Oman and the World: The Emergence of an Independent Foreign Policy* (Santa Monica, CA: Rand Corporation, 1995), 182.
2. Daniel Yergin, *The Quest* (New York: Penguin Press, 2011), 67.
3. Daniel Cuff, "Oil Trader a Big Winner in Atlantic Sale to Sun," *The New York Times*, published July 7, 1988, http://www.nytimes.com/1988/07/07/business/business-people-oil-trader-a-big-winner-in-atlantic-sale-to-sun.html.
4. "Deuss Handed Suspended Sentence for Banking Crimes," *The Royal Gazette*, published May 28, 2012, http://www.royalgazette.com/article/20120528/NEWS02/705289937.
5. "Deuss Receives 6 Month Suspended Sentence," *The Bermuda News*, published May 26, 2012, http://bernews.com/2012/05/six-month-suspended-sentence-for-john-deuss/.
6. Gareth Finighan, "Deuss Cuts a $47m Deal," *The Royal Gazette*, published August 3, 2013, http://www.royalgazette.com/article/20130803/NEWS03/130809982.
7. Steve LeVine, *The Oil and the Glory: The Pursuit of Empire and Fortune on the Caspian Sea* (New York: Random House, 2007), 191–94.
8. Arifa Alieva, "Azerbaijan Report: November 2, 2001," *Radio Free Europe/Radio Liberty*, November 2, 2001, http://www.rferl.org/content/article/1340906.html. In that larger report, the story appears in the fifth paragraph under the heading "Press Review," which summarizes key points from an newspaper article Ulker Ismailgizi wrote for the newspaper *Azadlig* about Marat Manafov.

Chapter 6

1. Paul Quinn-Judge, "Shevardnadze the Survivor," *The Washington Post*, published March 19, 2006, http://www.washingtonpost.com/wp-dyn/content/article/2006/03/17/AR2006031702090.html.

2. Marianne Lavelle and Amanda Rivkin, "Pictures: At Five Years Old, BTC Pipeline Moves Oil, Culture," *National Geographic*, published June 10, 2011, http://news.nationalgeographic.com/news/energy/2011/06/pictures/110608-baku-tbilisi-ceyhan-oil-pipeline/.
3. "Baku-Tbilisi-Ceyhan (BTC) Caspian Pipeline," *Hydrocarbons-Technology*, accessed June 19, 2014, http://www.hydrocarbons-technology.com/projects/bp/.

Chapter 7
1. "Korean Airlines Flight Shot Down by Soviet Union," *History.com*, accessed April 27, 2014, http://www.history.com/this-day-in-history/korean-airlines-flight-shot-down-by-soviet-union.
2. Thom Patterson, "The downing of Flight 007: 30 years later, a Cold War tragedy still seems surreal," *CNN*, published August 31, 2013, http://edition.cnn.com/2013/08/31/us/kal-fight-007-anniversary/.

Chapter 8
1. "Russia—The Family," *Country Studies (Russia)*, accessed February 10, 2014, http://countrystudies.us/russia/50.htm.
2. "Felix Dzerzhinsky," *History Learning Site*, published January 2012, http://www.historylearningsite.co.uk/felix_dzerzhinsky.htm.
3. Oleg Liakhovich, "Will Iron Felix Make a Return?" *The Moscow News* (#7, 2007), accessed February 10, 2014, https://sites.google.com/site/liakhovich/Index/felix.
4. Julia Rubin (AP), "Murder in Moscow Ends Dream of an American Entrepreneur," *Los Angeles Times*, published November 24, 1996, http://articles.latimes.com/1996-11-24/news/mn-2439_1_american-entrepreneur.
5. "Cathedral of Christ the Savior, Moscow," sacred-destinations.com, accessed February 10, 2014, http://www.sacred-destinations.com/russia/moscow-cathedral-of-christ-the-savior.

Chapter 9
1. "Khanty-Mansiysk Monthly Climate Average, Russia," *World Weather Online*, accessed February 10, 2014, http://www.worldweatheronline.com/Khanty-Mansiysk-weather-averages/Khanty-Mansiy/RU.aspx.
2. "Timeline: The rise and fall of Yukos," *BBC News*, last updated May 31, 2005, http://news.bbc.co.uk/2/hi/business/4041551.stm.
3. Richard Sakwa, *Putin and the Oligarch: The Khodorkovsky-Yukos Affair* (New York: I.B. Tauris & Co. Ltd, 2014), Amazon Books, accessed April 11, 2015, 12, http://www.amazon.com/Putin-Oligarchs-The-Khodorkovsky-Yukos-Affair/dp/1780764596#reader_1780764596.
4. "Rich in Russia—How to Make a Billion Dollars," *PBS*, published October 2003, https://www.pbs.org/frontlineworld/stories/moscow/khodorkovsky.html.

5. Thane Gustafson, *Wheel of Fortune: The Battle for Oil and Power in Russia* Cumberland, RI: Harvard University Press, 2012, 137–40.
6. Seth Mandel, "ExxonMobil's Role in Oil Tycoon's Arrest," *Commentary Magazine*, published May 1, 2012, http://www.commentarymagazine.com/2012/05/01/exxonmobil-role-in-khodorkovsky-arrest/.
7. "List of World's Richest People," *Forbes*, accessed February 11, 2014, http://www.forbes.com/2004/02/25/bill04land.html.

Chapter 10

1. "Litvinenko Murder: Coroner 'said Russia could be involved,'" *BBC*, published January 21, 2014, http://www.bbc.co.uk/news/uk-25824904.
2. Ann Curry, "Who Killed Alexander Litvinenko," *NBC News*, last updated July 17, 2007, http://www.nbcnews.com/id/17332541/ns/dateline_nbc-last_days_of_a_secret_agent/t/who-killed-alexander-litvinenko/#.UvpWlbmYapo.
3. Ibid.
4. Caroline DiCarlo, "Polonium-210 and the Assassination of Alexander Litvinenko," *Forensic Magazine*, published June 1, 2009, http://www.forensicmag.com/articles/2009/06/polonium-210-and-assassination-alexander-litvinenko.
5. Curry, "Who Killed Alexander Litvinenko."
6. DiCarlo, "Polonium-210."
7. Gen Roessler, "Why ^{210}Po?" *Health Physics News* (Volume XXXV Number 2), published February 2007, https://hps.org/documents/polonium_210_story.pdf.
8. Curry, "Who Killed Alexander Litvinenko."
9. Jason Bennetto, "Litvinenko inquiry closes in on suspected killers," *The Independent*, published January 6, 2007, http://www.independent.co.uk/news/uk/crime/litvinenko-inquiry-closes-in-on-suspectedkillers-430949.html.
10. Jane Croft, "UK High Court Hears Case for Public Inquiry into Litvinenko's Death," *Financial Times*, published January 21, 2014, http://www.ft.com/cms/s/0/abb2dad4-82ad-11e3-9d7e-00144feab7de.html#axzz3VzXkIxBX.
11. Richard Galpin, "Litvinenko Inquiry: Key Suspect Kovtun Aims to Clear His Name," *BBC*, published March 22, 2015, http://www.bbc.com/news/world-europe-32012759.

Chapter 11

1. "Literacy Rate, Adult Total," *World Bank*, accessed February 11, 2014, http://data.worldbank.org/indicator/SE.ADT.LITR.ZS?page=4.
2. "Vietnam Population," *Index Mundi*, accessed February 11, 2014, http://www.indexmundi.com/vietnam/population.html.
3. Christopher R. Cox, "The Storied History of Caravelle Hotel in Ho Chi Minh City," *DestinAsian*, published May 2, 2009, http://www.destinasian.com/countries/east-southeast-asia/vietnam/caravelle-hotel/.

Chapter 12

1. Kinette Lopez, "Study: Beijing's Air Pollution is Shaving Up to 16 Years off Chinese People's Lives," *Business Insider*, published January 21, 2014, http://www.business insider.com/beijing-air-quality-hits-life-expectancy-2014-1.
2. Liu Shibai, "A Great Revolution: Developing a Socialist Market Economic System," *Qiushi Journal (English Edition)*, last updated September 19, 2011, http://english.qstheory.cn/economics/201109/t20110924_112461.htm.
3. Patrick E. Tyler, "Deng Xiaoping: A Political Wizard Who Put China on Capitalist Road," *New York Times*, published February 20, 1997, http://www.nytimes.com/learning/general/onthisday/bday/0822.html.

Chapter 13

1. "SARS: Timeline of an Outbreak," *WebMD*, accessed February 25, 2014, http://www.webmd.com/lung/news/20030411/sars-timeline-of-outbreak.

Chapter 15

1. "Endangered Languages of Papua New Guinea," *Documenting Endangered Languages of the Pacific (DELP) at the University of Sydney*, accessed April 2, 2015, http://sydney.edu.au/arts/research_projects/delp/png-languages.php.
2. Brent Swancer, "Lost Lands, Cannibals, and the Mysterious Disappearance of Michael Rockefeller," *Mysterious Universe*, published March 23, 2015, http://mysteriousuniverse.org/2015/03/lost-lands-cannibals-and-the-mysterious-disappearance-of-michael-rockefeller/.
3. Sam Putnam, "Journeys and Reflections: 25 Years of the Michael C. Rockefeller Memorial Fellowship," *The Michael C. Rockefeller Memorial Fellowship*, accessed February 25, 2014, http://uraf.harvard.edu/michael-c-rockefeller-memorial-fellowship.
4. Swancer, "Lost Lands, and the Mysterious Disappearance of Michael Rockefeller."
5. Ibid.
6. Richard B. Stolley, "So Bad Even the Bloody Trees Can't Stand Up," *Life Magazine*, December 1, 1961.
7. "The Disappearance of Michael Rockefeller," *The Papua Heritage Foundation (PACE)*, accessed February 25, 2014, http://www.papuaerfgoed.org/en/The_disappearance_of_Michael_Rockefeller.
8. Swancer, "Lost Lands, Cannibals."
9. Marc Hoover, "Author Reveals Details About 1961 Disappearance of Michael Rockefeller," *Examiner.com*, published February 23, 2014, http://www.examiner.com/article/author-reveals-details-about-1961-disappearance-of-michael-rockefeller.
10. Bill Gifford, " 'Savage Harvest: A Tale of Cannibals, Colonialism, and Michael Rockefeller's Tragic Quest for Primitive Art' by Carl Hoffman," *The Washington Post*, published March 21, 2014, http://www.washingtonpost.com/opinions/savage-harvest-a-tale-of-cannibals-colonialism-and-michael-rockefellers-tragic-quest-for-primitive-art--by-carl-hoffman/2014/03/21/70f6b746-a934-11e3-8d62-419db477a0e6_story.html.

11. Ibid.
12. Carl Hoffman, *Savage Harvest: A Tale of Cannibals, Colonialism, and Michael Rockefeller's Tragic Quest for Primitive Art* (New York: HarperCollins, 2014).

Chapter 16
1. Al Jazeera Correspondent, "Timeline: The Marcos Regime," *Al Jazeera*, published September 22, 2011, http://www.aljazeera.com/programmes/aljazeera correspondent/2011/09/201192082718752761.html.
2. Kate McGeown, "What Happened to the Marcos Fortune?" *BBC*, published January 25, 2013, http://www.bbc.com/news/world-asia-21022457.
3. Kallie Szczepanski, "Ferdinand Marcos," *About.com*, accessed May 1, 2014, http://asianhistory.about.com/od/profilesofasianleaders/p/fmarcosbio.htm.
4. "Hôtel de Paris Monte-Carlo," *The Leading Hotels of the World*, accessed February 26, 2014, http://www.lhw.com/hotel/Hotel-de-Paris-Monte-Carlo-Monte-Carlo-Monaco.

Chapter 17
1. AIPN Model International Dispute Resolution Agreement (2004), *Association of International Petroleum Negotiators*.
2. Sallie L. Gaines, "Amoco, Argentine Partner Map Southern Offensive," *Chicago Tribune*, published September 6, 1997, http://articles.chicagotribune.com/1997-09-06/business/9709060160_1_southern-cone-amoco-corp-bridas.

Chapter 18
1. David Work, "James Wilson Vanderbeek Memorial," *AAPG Non-Technical and Memorials, AAPG Bulletin*, June 1994, accessed February 27, 2014, http://archives.datapages.com/data/bull_memorials/078/078006/pdfs/1010.html.
2. "2013 Data Book," *BG Group*, accessed February 27, 2014, 5, http://files.the-group.net/library/bggroup/files/pdf_433.pdf.
3. "Trinidad Marks 1990 Coup Attempt," *BBC*, published July 27, 2010, http://www.bbc.com/news/world-latin-america-10774647.
4. David McFadden, "Trinidad Fact-Finding Panel Taking Fresh Look at 1990 Coup Attempt by Radical Muslim Group," *Fox News (AP)*, published May 28, 2013, http://www.foxnews.com/world/2013/05/28trinidad-fact-finding-panel-taking-fresh-look-at-10-coup-attempt-by-radical/.
5. Jada Loutoo, "24 Persons Died in 1990 Attempted Coup," *Trinidad and Tobago Newsday*, published May 22, 2013, http://www.newsday.co.tt/crime_and_court/0,178036.html.
6. Afra Raymond, "Three Myths About Corruption," *TED Talks*, accessed March 3, 2014, http://www.ted.com/talks/afra_raymond_three_myths_about_corruption.html.

Chapter 19

1. "US Overtakes Saudi Arabia and Russia as Largest Oil Producer," *Institute for Energy Research*, published July 10, 2014, http://instituteforenergyresearch.org/analysis/u-s-overtakes-saudi-arabia-russia-worlds-biggest-oil-producer/.
2. Trevis Team, contributor, "US Oil Rig Count Trends Lower, Shouldn't Have Major Effect," *Forbes*, published March 17, 2015, http://www.forbes.com/sites/greatspeculations/2015/03/17/u-s-oil-rig-count-trends-lower-shouldnt-have-major-effect/.
3. "Baku-Tbilisi-Ceyhan (BTC) Caspian Pipeline," *Hydrocarbons Technology*, published March 16, 2015, http://www.hydrocarbons-technology.com/projects/bp/.
4. Hella Pick, "Eduard Shevardnadze Obituary," *The Guardian*, published July 7, 2014, http://www.theguardian.com/world/2014/jul/07/eduard-shevardnadze.

About the Author

WILLIAM A. YOUNG IS AN EXPERIENCED INTERNATIONAL negotiator. As an executive with Amoco, BP, Burlington Resources, ConocoPhillips, and BG Group, Young traveled to locations as diverse as Russia, China, Azerbaijan, and Argentina in pursuit of quality energy deals during his more than thirty-year career. He led or participated in negotiations on a number of important projects, including the Azeri-Chirag-Guneshli field development in the Azerbaijan sector of the Caspian Sea and the Baku-Tblisi-Ceyhan export pipeline. Young collaborated with the Harvard Business School on a case study concerning the financing of those ventures.

Young holds a BS in mathematics from Duke University and an MBA in finance from the Wharton School. A military veteran, Mr. Young served as a lieutenant in the US Navy Supply Corps. He currently lives in North Carolina where he maintains a small consulting practice (WAY Energy Consulting LLC), which provides commercial and negotiations advice and training to the international energy industry.

CPSIA information can be obtained at www.ICGtesting.com
Printed in the USA
LVOW07s1956240915

455582LV00007B/850/P